Introduction to the Principles of Materials Evaluation

Introduction to the Principles of Materials Evaluation

David C. Jiles

Wolfson Centre for Magnetics
Institute for Advanced Materials and Energy Systems
Cardiff University
Cardiff, U.K.

CRC Press
Taylor & Francis Group
Boca Raton London New York

CRC Press is an imprint of the
Taylor & Francis Group, an informa business

CRC Press
Taylor & Francis Group
6000 Broken Sound Parkway NW, Suite 300
Boca Raton, FL 33487-2742

© 2008 by Taylor & Francis Group, LLC
CRC Press is an imprint of Taylor & Francis Group, an Informa business

No claim to original U.S. Government works
Printed in the United States of America on acid-free paper
10 9 8 7 6 5 4 3 2 1

International Standard Book Number-13: 978-0-8493-7392-3 (Hardcover)

Visit the Taylor & Francis Web site at
http://www.taylorandfrancis.com

and the CRC Press Web site at
http://www.crcpress.com

Table of Contents

Preface

This book was developed from an introductory undergraduate course in nondestructive testing and materials evaluation that has been offered for many years at Iowa State University. It deals with the basic concepts of materials evaluation and adopts the viewpoint that the underlying principle behind all materials evaluation methods is the interaction of energy of some type with a material. In other words, in order to find out anything about a material, we need to interact with it in some way, and this must come in the form of one or more types of energy. The condition of the material can then be deduced from the resulting relationship between the form of the energy input and the form of the output. This means that materials evaluation techniques can be classified systematically in terms of the main divisions of classical phenomenology: mechanics, heat, light, sound, electricity, magnetism, and radiation.

The book begins by considering the various physical properties of materials that may be of interest for materials evaluation and the means for determining these properties. We look at the various types of energy: mechanical, acoustic, thermal, optical, electrical, magnetic, and radiative. We then study how each of these types of energy provides the basis for measurements that can be made on any material. The results of these measurements are simply relationships between the input energy and the output energy. These relationships are determined by the properties and condition of the material and so, by inference, can be used to evaluate the condition of the material. The role of the material in determining the relationship through its response to the energy is therefore central to our understanding. The interpretation of the results is usually based on empirical knowledge or on models, which comprise mathematical equations with adjustable parameters that can be used to describe practical situations and, less often, on first principles theories, which rely more on basic physical laws. Finally, we look at technological applications, including the important considerations of destructive vs. nondestructive testing for flaws, the concept of materials characterization in which the intrinsic materials properties are studied before flaws appear, the need for *in situ* measurements and, finally, the recurring issue of plant life extension and retirement for cause.

The underlying concept is to examine the physical bases for materials evaluation, rather than just presenting the subject as a set of empirical techniques. This is achieved by emphasizing common principles in materials evaluation procedures and by looking at the fundamental physics of materials evaluation in terms of energy interacting with a material in a controlled way. This can be approached by considering that the relationship between energy input and output is determined by the material condition alone. Therefore, any information that can be extracted about the condition of the material comes from these relationships. In materials evaluation, the task becomes understanding how we do this in various cases and extracting what is common in each of these cases.

David Jiles

Acknowledgments

I am grateful to the authors and publishers for permission to reproduce the following figures that appear in this book:

Figure 2.3 J.R. Davis (Editor), *Metals Handbook*, Desk Edition, 2nd Edition, ASM International, Materials Park, Ohio, 1998.

Figure 7.10 J.R. Davis, (Editor), *ASM Handbook*, Desk Edition, 2nd Edition, ASM International, Materials Park, Ohio, 1998.

Figure 7.11 Photograph by Neal Boenzi. Reprinted with permission from the *New York Times*.

Figure 8.15 H.J. Rindorf, Acoustic Emission Source Location in Theory and in Practice, *Brüel & Kjær Technical Review*, No. 2, 1981.

Figure 8.16 H.J. Rindorf, Acoustic Emission Source Location in Theory and in Practice, *Brüel & Kjær Technical Review*, No. 2, 1981.

Figure 10.10 W. Lord, *IEEE Transactions on Magnetics*, 19, 2437, 1983.

Figure 11.4 L. Cartz, *Nondestructive Testing*, ASM International, Materials Park, Ohio, 1995.

Figure 11.6 *Nondestructive Testing*, Vol. 03.03, Section 3, ASTM Standards, American Society for Testing and Materials, Philadelphia, Pennsylvania.

Figure 11.7 *Nondestructive Testing*, Vol. 03.03, Section 3, ASTM Standards, American Society for Testing and Materials, Philadelphia, Pennsylvania.

Figure 11.8 *Nondestructive Testing*, Vol. 03.03, Section 3, ASTM Standards, American Society for Testing and Materials, Philadelphia, Pennsylvania.

Figure 14.5 S. N. Rajesh, L. Udpa, and S. S. Udpa, Review of Progress in Quantitative Nondestructive Evaluation, D. O. Thompson and D. E. Chimenti, Editors, Vol. 12B, 1993, pp. 2365–2372, Plenum, New York.

Figure 14.6 S.N. Rajesh, L. Udpa, and S.S. Udpa, *IEEE Trans Mag.* 29, 1857, 1993.

Figure 15.4 R.H. Burkel et al., U.S. Department of Transportation, Federal Aviation Administration Report No. DOT/FAA/AR-01/96, September 2002.

1 Introduction

This chapter provides an overview of some of the considerations that go into materials evaluation and nondestructive testing. It includes the concept of materials' lifetime, whereby the degradation and eventual replacement or failure of materials are considered as normal, and a discussion of factors that can cause failure of materials. In addition to a summary of some of the most basic inspection techniques, it describes the considerations that arise when a material is found to be flawed, particularly whether the material should be replaced or whether continued operation is possible. The chapter forms a foundation for the ideas that are presented in the following chapters of the book.

1.1 FUNDAMENTALS OF MATERIALS EVALUATION AND THE CONCEPT OF LIFETIME OF MATERIALS

Perhaps the most important concept in materials evaluation and nondestructive testing is that materials change with time and that these changes need to be considered as normal. This concept is somewhat different from the traditional viewpoint of materials science in which a material is fabricated and that is the end of the process, almost as if the material will now remain the same indefinitely, unless something unexpected happens to it.

In practice, when a material is fabricated for a particular application, it is then used under certain operating conditions and, after service exposure, the material is not the same as when it began life. The changes usually arise in the form of degradation of the materials' properties and, if this process continues, then it can eventually lead to failure. It is therefore desirable to monitor the degradation of the material, if for no other reason than to identify when failure is likely to occur.

1.1.1 EFFECTS OF DIFFERENT FORMS OF ENERGY ON MATERIALS

Structural changes occur in a material as a result of prolonged exposure to high levels of different forms of energy. These changes can include residual stress (increased dislocation density), embrittlement (decrease in toughness), fatigue (tendency for failure after exposure to repeated cycling of applied stress), creep (slow flow of material under applied stress and elevated temperature leading to voids and cavities in the material), and radiation damage (accumulation of defects or migration of chemical species resulting from interactions with radiation).

1.1.2 FACTORS THAT CAN CAUSE FAILURE OF A MATERIAL

We now consider those external factors that can cause a material to fail. These almost always have a mechanical component and, in some cases, the mechanical component acts alone; in other cases, the mechanical component acts together with other factors, such as temperature, chemical action, or radiation. Some of the main failure mechanisms are: (1) fatigue, (2) creep, (3) corrosion, (4) oxidation, (5) thermal embrittlement, and (6) radiation embrittlement.

Some failures arise as a combination of the preceding factors.

1.2 TESTING

Materials testing can be either destructive or nondestructive. Traditional destructive tests used by materials scientists include the following:

1.2.1 MACROEXAMINATION

This is a traditional destructive examination, which involves cutting and polishing a material to look for macroscopic flaws and inhomogeneities. After polishing the material, it is often etched to emphasize metallurgical differences.

1.2.2 MICROEXAMINATION (MICROGRAPHS)

This consists of cutting and polishing a material, followed by examination under a microscope at about 500X, or using a scanning electron microscope at higher magnifications. The surface is usually photographed (for later examination) to produce a micrograph, which allows documentation and examination of the metallurgical microstructure.

1.2.3 COMPARISON OF VISUAL INSPECTION WITH OTHER METHODS

Table 1.1 and Table 1.2 show how the various nondestructive evaluation (NDE) techniques compare in terms of equipment costs, speed of inspection, restrictions on geometry of test part, types of defects, required skill of operator, portability of equipment, ease of automation, and the general types of problems that each can address.

1.2.4 UNASSISTED VISUAL INSPECTION

Visual inspection is the oldest and most straightforward method of NDE. In essence, it just amounts to investigating whether there are any obvious signs of degradation of a material on its surface. Many techniques have been developed to enhance the capability of visual inspection. These include the use of lenses to increase the size of images, dye penetrants (including ultraviolet dye penetrant techniques), optical and electron microscopy, and software-assisted methods, such as digital image enhancement methods, to improve detectability of flaws.

TABLE 1.1
Applications of Different NDE Techniques

	Test Method				
	Ultrasonics	**X-Ray**	**Eddy Current**	**Magnetic Particle**	**Liquid Penetrant**
Capital cost	Medium to high	High	Low to medium	Medium	Low
Consumable cost	Very low	High	Low	Medium	Medium
Time of results	Immediate	Delayed	Immediate	Short delay	Short delay
Effect of geometry	Important	Important	Important	Not too important	Not too Important
Access problems	Important	Important	Important	Important	Important
Type of defect	Internal	Most	External	External	Surface breaking
Relative sensitivity	High	Medium	High	Low	Low
Formal record	Expensive	Standard	Expensive	Unusual	Unusual
Operator skill	High	High	Medium	Low	Low
Operator training	Important	Important	Important	Important	Low
Training needs	High	High	Medium	Low	Low
Portability of equipment	High	Low	High to medium	High to medium	High
Dependent on material composition	Very	Quite	Very	Magnetic only	Little
Ability to automate	Good	Fair	Good	Fair	Fair
Capabilities	Thickness gauging; some composition testing	Thickness gauging	Thickness gauging; grade sorting	Defects only	Defects only

In the case of unassisted visual inspection, it is possible, for example, to detect deterioration of welds, problems with corrosion and pitting of ferrous materials, and larger cracks.

1.2.5 ASSISTED VISUAL OBSERVATION

There are various means of enhancing the visual inspection method, which include making micrographs using either optical or electron microscopy. Under conditions of assisted observation, such as a sample that has been etched and polished and then studied under a microscope at magnifications of typically about 500 ×, the grains of the material can usually be seen so that examination of the microstructure is possible, including grain sizes, grain boundaries, and the presence of second phases. Problems such as intergranular stress corrosion cracking can be seen. Some examples of the kinds of features that can be seen under these conditions are shown in the following figures (Figure 1.1 to Figure 1.3).

Electron micrographs using scanning electron microscopy, with magnifications of typically 1000X, can provide higher spatial resolution than optical micrographs.

TABLE 1.2
Comparison of Different NDE Techniques

Method	Characteristics Detected	Advantages	Limitations	Example of Use
Ultrasonics	Changes in acoustic impedance caused by cracks, nonbonds, inclusions, or interfaces	Can penetrate thick materials; excellent for crack detection; can be automated	Normally requires coupling to material either by contact to surface or immersion in a fluid, such as water. Surface needs to be smooth	Adhesive assemblies for bond integrity; laminations; hydrogen cracking
Radiography	Changes in density from voids, inclusions, material variations; placement of internal parts	Can be used to inspect wide range of materials and thicknesses; versatile; film provides record of inspection	Radiation safety requires precautions; expensive; detection of cracks can be difficult unless perpendicular to x-ray film	Pipeline welds for penetration, inclusions, voids; internal defects in castings
Visual-optical	Surface characteristics such as finish, scratches, cracks, or color; strain in transparent materials; corrosion	Often convenient; can be automated	Can be applied only to surfaces, through surface openings, or to transparent material	Paper, wood, or metal for surface finish and uniformity
Eddy current	Changes in electrical conductivity caused by material variations, cracks, voids, or inclusions	Readily automated; moderate cost	Limited to electrically conducting materials; limited penetration depth	Heat exchanger tubes for wall thinning and cracks
Liquid penetrant	Surface openings due to cracks, porosity, seams, or folds	Inexpensive, easy to use; readily portable; sensitive to small surface flaws	Flaw must be open to surface; not useful on porous materials or rough surfaces	Turbine blades for surface cracks or porosity; grinding cracks
Magnetic particles	Leakage magnetic flux caused by surface or near-surface cracks, voids, inclusions, material or geometry changes	Inexpensive or moderate cost, sensitive both to surface and near-surface flaws	Limited to ferromagnetic material; surface preparation and post-inspection demagnetization may be required	Railroad wheels for cracks; large castings

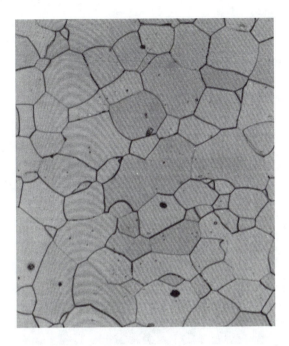

FIGURE 1.1. Polished and etched grains of a sample of iron as they appear under an optical microscope.

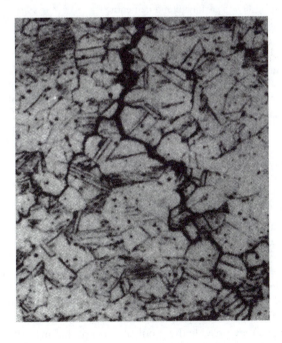

FIGURE 1.2 Optical micrographs showing intergranular stress corrosion cracking in brass.

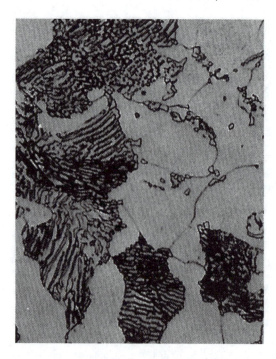

FIGURE 1.3 Optical micrograph showing the pearlitic structure of the grains, which consist of alternating layers of ferrite and cementite.

1.3 VARIOUS METHODS IN MATERIALS EVALUATION

In most cases, the evaluation of materials is through measurement of the physical properties of materials. This involves utilizing the interaction of different forms of energy with materials. These could be one or more of the following:

- Mechanical
- Acoustic
- Thermal
- Optical
- Electrical
- Magnetic
- Radiative

Under well-controlled laboratory conditions, inspection methods have few constraints, and so the instrumentation can be designed to get the best possible measurement of the materials' properties of interest. These measurements can be important, because they provide the proof of concept that measurement of certain physical properties can be linked to changes in the condition of materials. Without such information, the development of practical fieldable inspection methods is much more difficult, so the proof of concept studies have a vital role to play. Certainly, if the laboratory test shows no success, there is little chance that a practical fieldable method based on the same concept will succeed. Practical implementation of the methods, however, usually requires some compromises because, in industrial environments, there are usually additional constraints on the measurement, which determine that the equipment is not necessarily capable of getting the best possible measurement. Loss of signal, accuracy, or precision is the usual result. However, this does not necessarily mean that the method is unsuitable, just that there will be some additional factors bearing on the measured result.

1.3.1 Concept for Nondestructive Evaluation

When a test material needs to be interrogated to assess its condition, it should be affected or impacted appropriately (energy input) and its response (energy output) studied. The relationship between the input and the output gives information about the state of the material. The general concept for NDE of materials is that it is not just the output alone that gives information about the material, but rather the relationship between the output and input. A variety of different types of interrogation methods are available, classified principally by the type of energy used. These can be mechanical, acoustic, thermal, optical, electrical, magnetic, or radiative, depending on the information required from the material.

1.3.2 Techniques for Nondestructive Assessment

Generally, we need to find some way of interrogating materials. This means interaction of various forms of energy, such as the following, with the material under examination. The list in section 1.3 enables us to classify the main NDE methods.

1.3.2.1 Major Methods

The major methods are as follows:

- Acoustic (ultrasonics)
- Mechanical
- Electrical (eddy currents)
- Radiative (x-rays)
- Fluorescent dye penetrants

1.3.2.2 Minor Methods

The minor methods are the following:

- Magnetic
- Optical
- Thermal

1.3.3 REMEDIAL ACTION

It is not always necessary to replace every damaged part that is identified. In general, the decision involves a more complex set of considerations, which eventually resolve to one of the following actions:

1. Replace
2. Repair
3. Continue service

Replace usually means simply removing the old part and putting a new one in its place. This will result in continued operation but, depending on the complexity of the part, could be expensive.

Repair could cover a wide range of possibilities, including welding, brazing, soldering, adhesive bonding, or, in other cases, the removal of some of the material — for example, by grinding of cracks to remove the damaged region.

Continue service means taking no action and should be considered if it can be demonstrated that the damaged part is not critical to the operation or that the damage is minimal enough that the likelihood of failure is low.

1.3.4 TERMINOLOGY

There is a wide range of terminology used in the evaluation of materials. Some of these seem to be almost indistinguishable, but each has its own distinctive meaning.

Quality control	Checking material as it is produced
Quality assurance	Checking material as it is received
Materials evaluation	Generic term for property measurements, including those for destructive as well as nondestructive purposes
Materials characterization	Focusing on intrinsic properties; may include destructive methods
Nondestructive testing	Usually relates to failure processes, detection of flaws, etc.; often qualitative in nature
Nondestructive evaluation	Relates to failure processes; more quantitative than nondestructive testing
Destructive testing	This is a form of testing in which a representative set of samples is drawn from a larger group and the representative group is tested to destruction, to determine their properties and condition; the assumption is made that the remaining samples are the same as the group that were tested so that the information obtained in the destructive tests can be applied to the remaining undamaged samples

FURTHER READING

Bray, D.E. and Stanley, R.K., *Nodestructive Evaluation: A Tool in Design, Manufacturing and Service*, CRC Press, Boca Raton, FL, 1997.

Cartz, L., *Nondestructive Testing*, ASM International, Ohio, 1995.

Davis, J.R. (Ed.), *Metals Handbook — Desk Edition*, ASM, OH, 1998, pp.1254–1307.

Davis, J.R. (Ed.), *Metals Handbook — Desk Edition*, ASM, OH, 1998, pp.1308–1355.

Eberhart, M.E., *Why Things Break*, Three Rivers Press, New York, 2003.

Flewitt, P.E.J. and Wild, R.K., *Physical Methods for Materials Characterization*, IOP, Bristol, 1994.

Halmshaw, R., Nondestructive Testing, 2nd ed., Edward Arnold Publishers, London, 1991.

Halmshaw, R., *Mathematics and Formulae in NDT*, 2nd ed., British Institute of Non destructive Testing, Northampton, 1993.

Hayward, G.P., *Introduction to Nondestructive Testing*, American Society for Quality Control, MI, 1978.

Hellier, C.J., *Handbook of Nondestructive Evaluation*, McGraw-Hill, New York, 2001.

Iddings, F.A., NDT: What, where, why, and when, *Mater Eval* 56, 505, April 1998.

Mitra, A., Parida, N., Bhattacharya, D.K., and Goswami, N.G. (Eds.), *Nondestructive Evaluation of Materials*, NML, Jamshedpur, India, 1997.

NDE Website www.ndt-ed.org.

Ness, S. and Sherlock, C.N., *Nondestructive Testing Overview*, ASNT, Columbus, 1996.

O'Connor, T.V. and Francen, C., Why do NDT? *Mater Eval* 57, 705, July 1999.

Singh, G.P., Schmalzel, J.L., and Udpa, S.S., Fundamentals of data acquisition for non-destructive evaluation, *Mater Eval* 48, 1341, 1990.

Grandt, A.F., *Fundamentals of Structural Integrity: Damage Tolerant Design and Nondestructive Evaluation*, John Wiley & Sons, Hoboken, NJ, 2004.

2 Mechanical Properties of Materials

The mechanical condition of materials is perhaps the most important single consideration in failure, and is of primary concern in nondestructive evaluation. Therefore, our investigation begins with the study of mechanical properties. This includes elastic and inelastic properties of materials. Stress/strain curves give us very basic information on the mechanical condition of a material. Hooke's law, although useful in many cases, is an approximation only and is insufficient for larger deformations. Even when describing linear elastic behavior in three dimensions, it needs to be generalized to include transverse strains. Extension of the mechanical behavior to cover nonlinear deformation is relevant, because failure often involves high levels of stress that go beyond the linear reversible regime. Hardness measurements are discussed as a means of determining the resistance of a material to plastic deformation.

2.1 EFFECTS OF STRESS ON A MATERIAL

If we list the ways that stress can affect a material, we find that it can cause strain, which can be elastic deformation for low stress levels, and plastic deformation at higher stress levels. Failure can be caused by a single application of stress, when it exceeds the tensile strength, whereby the material is effectively stretched until it fails. Failure can be caused by repeated application of stress (cyclic stress) at levels below the tensile strength through a process known as *fatigue*. Fatigue itself can be subdivided into two categories, known as low-cycle fatigue (for stresses that exceed the yield strength) and high-cycle fatigue (for stresses that do not exceed the yield strength). Failure can also occur as a result of the application of stress at elevated temperature. This latter process is known as *creep* and amounts to a plastic flow of the material, which can often be slow at first (primary creep), then almost reduces to zero rate of strain over the "plateau region" (secondary creep), but finally increases in rate again as the material approaches failure (tertiary creep).

In addressing these effects of stress, we must consider which of these processes are reversible and which result in a permanent change in the material. Among those that result in permanent changes, a subset leads to failure. Although these modes of failure are different, in all cases cracks form in the material. These cracks grow and sometimes join together until, eventually, the material is no longer able to sustain the stress. This leads to failure. From the viewpoint of materials evaluation, the objective is to determine how much permanent damage has been done to a material and how close it is to failure. One way to achieve this is through mechanical testing.

2.1.1 Mechanical Testing

Different forms of mechanical testing include the following:

- Tensile testing
- Hardness testing
- Fatigue testing
- Creep testing
- Impact testing

Elastic moduli can be obtained from tensile tests and tell us about the deformation of material on the continuum scale, but this is really only applicable for small stresses. On the atomistic scale, the elastic moduli give a measure of the average strength of interatomic interactions. The elastic moduli are not constants, but can vary for a number of reasons. These can include mechanical damage, temperature, microstructural changes, plastic deformation, and so on. Therefore, measurement of elastic moduli and the study of the deviation from previously measured or expected values can be used for nondestructive evaluation because it can indicate changes in the mechanical condition of a material. The measurements of elastic moduli are subject to some limitations — particularly for high strains involving nonlinear deformations where Hooke's law no longer applies, or cases where cracks occur in the material, whereby a measurement of the elastic modulus without correcting for the change in load-bearing area can give erroneous results.

Other mechanical properties may be of interest for materials evaluation. These include the ultimate tensile strength and the fracture toughness. However, such properties are obtained from destructive tests performed on a selected representative group of samples, and it is then assumed that the properties of the untested materials are the same as those obtained with the representative group.

2.1.1.1 Destructive vs. Nondestructive Testing

It is worth asking at what point a nondestructive test becomes destructive. The ideal to be aimed at is the use of measurements that cause no disruption of the material. This can sometimes be impossible to achieve technically, or is otherwise impractical, because the cost of the nondestructive inspection may exceed the value of the test material resulting from the nondestructive examination. Therefore, in practice, the boundary between the two types of tests can get blurred with certain tests, such as the hardness test, that cause minimal damage when being used for nondestructive testing. In other cases, destructive tests are deliberately used on a representative sample of materials to determine the properties of the remaining untested materials.

2.1.2 The Stress–Strain Curve

The variation of strain with stress that is obtained from a tensile test produces a stress–strain curve [1]. This usually shows a linear variation of strain with stress

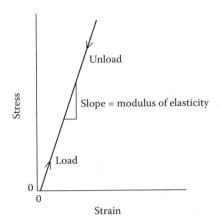

FIGURE 2.1 Variation of stress and strain on a material.

at low stress levels but becomes nonlinear at higher stress levels. The onset of the nonlinearity occurs as the deformation of the material changes from elastic strain (reversible) to plastic strain (irreversible). The following mechanical properties of materials can be obtained from the stress–strain curve:

- Young's (elastic) modulus
- Yield strength
- Tensile strength
- Ductility
- Toughness
- Resilience

The variation of strain with stress in the elastic range is shown in Figure 2.1. This is a straight-line graph passing through the origin and, from this, the elastic modulus can be obtained as the derivative of stress with respect to strain. The inverse of this is the elastic stiffness, the derivative of strain with respect to stress.

When the level of stress passes beyond the elastic range, the strain begins to vary more rapidly with stress as shown in Figure 2.2. This leads to a nonlinearity in the curve of stress vs. strain, which is the characteristic feature of plastic deformation. When the material is unloaded from the plastic regime, the strain does not completely recover. Instead, it is usual for the stress–strain locus to recoil along a linear trajectory parallel to the initial elastic deformation line. The strain that remains when the stress is reduced to zero is the plastic deformation, and the difference between this and the deformation under load is the elastic strain that is recovered as the load is removed.

A number of different mechanical properties can be defined from a stress–strain curve [2; pp. 275–289]. The ways in which the principal mechanical properties can be obtained from the stress–strain curve are shown in Figure 2.3.

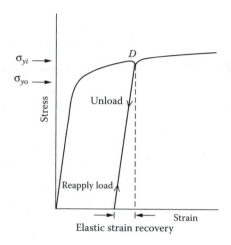

FIGURE 2.2 Elastic and plastic ranges of the stress–strain relationship.

2.1.3 Elastic Modulus

This is the coefficient in Hooke's law and is found from the slope of the stress–strain curve over the linear regime at low stress before plastic deformation occurs. There are several different elastic moduli that can be measured, depending on the nature of the stress and the resulting strain.

2.1.4 Yield Strength

This is the stress needed to cause plastic deformation (yielding) of the material under a unidirectional load. Because it is often difficult to establish exactly where yielding actually starts, an alternative definition is often used by engineers, and this is that the *yield strength* is the stress needed to induce a plastic deformation of 0.2%. This has the advantage of at least being quantifiable.

2.1.5 Plastic Deformation

On the macroscopic scale, when stress is removed from a material, if the strain does not completely recover, the resulting residual strain is called *plastic deformation*. On the atomistic scale, this occurs when crystal planes slip past each other without failure. This leads to nonlinear stress–strain behavior.

2.1.6 Tensile Strength

This is the stress needed to cause failure of the material under a unidirectional load. The true stress is the load divided by the actual perpendicular cross-sectional area. In many engineering applications, the engineering stress is used instead; this is the load divided by the original perpendicular cross-sectional area. The real cross-sectional area decreases because of necking and then cracking. Therefore, in the

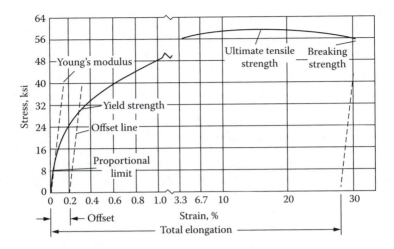

FIGURE 2.3 Different properties obtained from the stress–strain plot.

case of a graph of strain vs. engineering stress, there can be an apparent decrease in stress as the material comes close to failure, and this is depicted in Figure 2.4, where the tensile stress at failure appears to be less than the maximum applied stress. This, however, is merely an artifact, and if corrections are made for the reduction in cross-sectional area, it will be seen that the stress does not decrease and that the ultimate tensile strength represents the maximum stress.

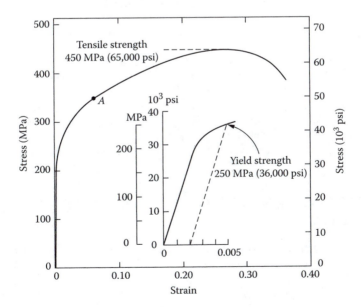

FIGURE 2.4 Plot of stress vs. strain for a specific material.

2.1.7 Ductility

Ductility of the material can been defined in either of two ways; either the linear ductility, which is the fractional change in length at the point of failure, or the area ductility, which is the fractional change in cross-sectional area at the point of failure. These two can be identical under certain conditions, but, in general, they will have different values.

2.1.8 Toughness and Resilience

These may sound similar, but they refer to different quantities obtained from the stress–strain curve. Resilience is the area under the stress–strain curve up to the yield point. It is, therefore, the amount of energy needed per unit volume to cause yielding of the material. Toughness, on the other hand, is the area under the stress–strain curve of the material up to the failure point, and so can be viewed as the energy per unit volume needed to cause failure.

Worked example

From the given figure (Figure 2.4), which shows the stress–strain behavior of a material, determine:

1. The elastic modulus
2. The yield strength (at 0.002 strain)
3. The maximum load sustainable by a specimen with 12.8-mm diameter
4. The change in length of a 250-mm rod at 345 MPa

Solution

1. From the slope of the graph, $E = d\sigma/de$: $E = 93.8$ GPa.
2. Using the 0.2% value to determine the yield strength, 0.002 strain occurs at 250 MPa.
3. Maximum load depends on the highest value of stress on the stress–strain curve. The reduction in area is sometimes taken into account, sometimes not. Without taking this into account, $F_{max} = 57,900$ N.
4. The change in length dl equals the product of strain e and original length l_o; dl $= e \cdot l_o = 0.06 \times 250 = 15$ mm.

2.2 STRESS–STRAIN RELATIONSHIPS AND ELASTIC PROPERTIES

A variety of elastic moduli are defined in terms of the ratios of different stresses to strains. Generally,

$$E = \frac{\sigma}{e}. \tag{2.1}$$

The reciprocal of this quantity is known as the *elastic compliance*,

$$S = \frac{e}{\sigma}. \tag{2.2}$$

In isotropic materials, the elastic properties can be completely specified by knowing two of the following quantities: Young's modulus, Poisson's ratio, bulk modulus, or shear modulus. Exact relations, therefore, exist between any three elastic moduli of a particular isotropic, homogeneous material. For example, shear modulus or bulk modulus can be expressed in terms of the Young's modulus and Poisson's ratio [3].

In three dimensions, we can define stresses and strains along different directions. This requires a matrix approach relating the various strains to the stresses through a stiffness or compliance matrix. Most of what is encountered in standard mathematical theory is just a generalized Hooke's law, which is an approximation that assumes linear reversible behavior (for example, when normal stresses are applied to a volume of material, as shown in Figure 2.5).

The number of independent elastic moduli increases in anisotropic and textured materials, because the elastic moduli are then different along different directions. The maximum number of independent elastic moduli is 21, which, in the lowest symmetry structures, comprises 3 longitudinal moduli, 3 shear moduli, and 15 transverse moduli.

2.2.1 LONGITUDINAL STRESS AND STRAIN: YOUNG'S MODULUS

Young's modulus is the ratio of longitudinal strain to longitudinal stress, where the directions of stress and strain are parallel

$$Y = \frac{\sigma_{//}}{e_{//}}. \tag{2.3}$$

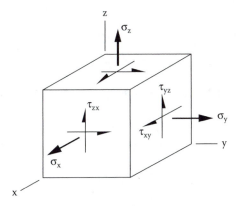

FIGURE 2.5 Normal (perpendicular) stress on a volume of material.

In three dimensions, the stresses and strains can be described in terms of components along the principal axes, x, y, and z. Conventionally, the stresses along the principal axes are denoted as σ_x, σ_y, and σ_z. The corresponding longitudinal strains are e_{xx}, e_{yy}, and e_{zz}. But, in addition, there are transverse strains e_{xy}, e_{yx}, e_{yz}, e_{zy}, e_{zx}, and e_{xz}; where e_{xy} denotes the strain along the x-axis due to a stress along the y-axis and so on.

2.2.2 TRANSVERSE STRAIN: POISSON'S RATIO

Transverse strain is not the same as shear strain. It is a longitudinal strain that arises as a result of a longitudinal stress in another direction perpendicular to the strain. The magnitude of the transverse strain is determined by Poisson's ratio. In a material that is isotropic and for which the volume is conserved under applied longitudinal stress, $\nu = 0.5$. In practice, the Poisson's ratio is below this value.

When a material is strained in one direction by a uniaxial stress, it usually suffers strain in the orthogonal directions. Poisson's ratio gives the amount of transverse strain relative to the parallel strain,

$$\nu = \frac{e_\perp}{e_{//}}.$$

(2.4)

With one or two notable exceptions, the application of a longitudinal tensile stress results in a transverse compressive strain. For conservation of volume, of course, ν must have a value of 0.5 but, in most cases, volume is not conserved, and has a value closer to 0.25 to 0.3. In steels, for example, $\nu \sim 0.3$.

These strains can then be expressed in terms of relevant stresses and elastic moduli,

$$e_{xx} = \sigma_x/Y$$

$$e_{yy} = \sigma_y/Y$$

(2.5)

$$e_{zz} = \sigma_z/Y.$$

There will also be transverse strains by Poisson's effect, so that

$$e_{xy} = -\nu\sigma_y/Y$$

$$e_{yz} = -\nu\sigma_z/Y$$

(2.6)

$$e_{zx} = -\nu\sigma_x/Y,$$

where e_{xy} is the strain along the x-axis resulting from a stress along the y-axis, and so on.

The total strains are therefore obtained by summing the various contributions along each axis, for example, $e_x = e_{xx} + e_{xy} + e_{xz}$, giving a generalized three-dimensional version of Hooke's law for linear deformations [4; p. 106],

$$e_x = \frac{1}{Y}(\sigma_x - \nu\sigma_y - \nu\sigma_z)$$

$$e_y = \frac{1}{Y}(\sigma_y - \nu\sigma_z - \nu\sigma_x) \qquad (2.7)$$

$$e_z = \frac{1}{Y}(\sigma_z - \nu\sigma_x - \nu\sigma_y).$$

2.2.3 Shear Stress and Strain: Shear Modulus

The rigidity modulus or shear modulus is the ratio of shear strain θ to shear stress σ

$$G = \frac{\sigma}{\theta}. \qquad (2.8)$$

Shear strain is measured as an angular deformation θ. In the simplest possible case, if we consider a shear force (or couple) F acting on the sides of a square sample, the shear stress is given by $\sigma = F/a^2$, where a^2 is the cross-sectional area over which the shear force operates,

$$G = \frac{F}{a^2\theta}. \qquad (2.9)$$

If we consider the strain along the diagonal, it is given by

$$e = \frac{\sigma}{Y} + \frac{\nu\sigma}{Y}. \qquad (2.10)$$

The net resolved force along the diagonal is therefore

$$F_{res} = 2F \cos 45 = \frac{2F}{\sqrt{2}}. \qquad (2.11)$$

The area of the plane perpendicular to stress is

$$A = a^2\sqrt{2}. \qquad (2.12)$$

Therefore, the resolved stress along the diagonal is $\frac{F}{a^2}$.
The strain due to this resolved stress is

$$= 2\left(\frac{F}{a^2 Y} + \frac{vF}{a^2 Y}\right) = \frac{2F(1+v)}{a^2 Y}. \tag{2.13}$$

Therefore, the shear modulus is related to the Young's modulus Y and the Poisson's ratio v by the equation [4, p. 138]

$$G = \frac{Y}{2(1+v)}. \tag{2.14}$$

Generalizing to three dimensions, the shear forces on the cube are shown in Figure 2.6.

In three dimensions, if we consider the shear stress on an elemental cube of material, then, to have equilibrium, all forces must balance.

$$\sigma_{xy} = \sigma_{yx}$$

$$\sigma_{yz} = \sigma_{zy} \tag{2.15}$$

$$\sigma_{zx} = \sigma_{xz}.$$

From the shear-strain expression of Hooke's law

$$e_{xy} = \sigma_{xy}/G$$

$$e_{yz} = \sigma_{yz}/G \tag{2.16}$$

$$e_{zx} = \sigma_{zx}/G.$$

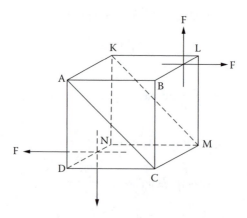

FIGURE 2.6 Three-dimensional shear forces on a cube.

There are no Poisson terms in these shear-strain equations; so they have a simpler form than that for the normal strains.

2.2.4 VOLUMETRIC STRAIN UNDER UNIAXIAL STRESS

The simplest case is that of uniaxial tensile stress applied to a uniform homogeneous material. If the initial unstrained volume is

$$V_i = xyz, \tag{2.17}$$

the final strained volume is

$$V_f = (x + dx)(y - dy)(z - dz), \tag{2.18}$$

which is approximately equal to

$$V_f = xyz + yzdx - xzdy - xydz; \tag{2.19}$$

and if $V_f = V_i$ it follows that

$$dx/x = dy/y + dz/z, \tag{2.20}$$

and because $dy/y = dx/x$, and $dz/z = dx/x$, it follows that for conservation of volume, $v = 0.5$.

WORKED EXAMPLE

A cylindrical steel rod, original diameter 12.8 mm, is tested to failure and has an engineering ultimate tensile strength of $\sigma_{UTS} = 460$ MPa. If its cross-sectional diameter at failure is 10.7 mm, find (1) the ductility, and (2) the true ultimate tensile strength.

Solution

Ductility is the percentage reduction in area. In this case, $\Delta A/A \times 100\% = \{(5.35)^2 - (6.4)^2\}/(6.4)^2 = -30\%$.

True stress is load/true area. In this case, the true ultimate tensile strength is $\sigma_{UTS} = F/A = 460$ MPa $\times (12.8/10.7)^2 = 658$ MPa.

2.2.5 VOLUMETRIC STRAIN UNDER HYDROSTATIC STRESS

The bulk elastic modulus is the ratio of volume strain dV/V to hydrostatic pressure P,

$$K = \frac{P}{dV/V}. \tag{2.21}$$

Consider the stresses on a parallelepiped in which the faces are not of equal area. The loads on each face are not identical if the stresses (pressures) on each face are identical. The relationship between stress and strain, in this case, is given by the bulk modulus.

To demonstrate the relationship between bulk modulus and other elastic moduli, such as Young's modulus, we consider a cube of unit dimensions compressed by a pressure P. The stress on each face is then

$$\sigma = P = F/A, \tag{2.22}$$

and each side of the cube strains by an amount e

$$e = -\sigma/Y + 2v\sigma/Y, \tag{2.23}$$

The volume strain is therefore

$$\frac{dV}{V} = 1 - \left(1 - \frac{\sigma}{Y} + \frac{2v\sigma}{Y}\right)^3, \tag{2.24}$$

and approximating this strain using only the first-order terms

$$\frac{dV}{V} = \frac{3\sigma(1-2v)}{Y}. \tag{2.25}$$

Because $K = \frac{\sigma}{dV/V}$, the bulk modulus is related to the Young's modulus Y and the Poisson's ratio v by the equation [4, p. 138]

$$K = \frac{Y}{3(1-2v)}. \tag{2.26}$$

2.2.6 TORSIONAL STRESS AND STRAIN

In defining the torsional strain in a rod, for example, it is important that when subjected to a uniform torque at the ends of the rod, the strain be the same throughout the rod. Therefore, the angle that measures this strain is angle ϕ in Figure 2.7. Note that the angle θ varies along the length of the rod and is a measure of displacement rather than strain.

Note also, that despite some ambiguity in common usage, the torsional stress is not the same as torque. Torque, also known as the turning moment, is measured in units of newton meters, and is the product of a force and the perpendicular distance over which it acts. Torsion, or torsional stress, is measured in newtons per square meter. The torsional stress can be calculated from the torque, normalizing it by the volume over which it acts.

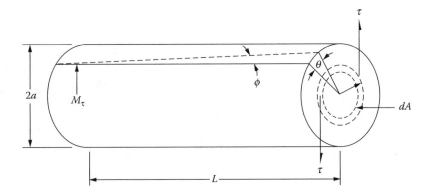

FIGURE 2.7 Illustration of torsional strain.

2.3 HARDNESS

Hardness is the resistance of material to permanent plastic deformation under a load. This property is then complementary to the elastic modulus of a material, which is a measure of its resistance to elastic deformation under the action of a stress. Because permanent plastic deformation only occurs beyond the yield strength, it is reasonable to expect that there is some relationship between hardness and yield strength.

There are several types of hardness tests, which can be categorized into three groups: indentation tests, scratch tests, and rebound tests [5]. The indentation tests, which are the most widely used measurement for hardness, include the Brinell, Rockwell, Vickers, and Knoop tests. There is one main scratch test, known as the Mohs test; and one main rebound test, which is the scleroscope test.

An example of an indentation test is shown schematically in Figure 2.8.

Each of the standard indentation tests has a different scale by which the hardness is quantified. There are exact mathematical formulae which are used to determine the hardness values on each of these different scales from the indentation data [6]. The values of the various type of hardness can be calculated from the measured indentation under controlled conditions by the following equations.

2.3.1 BRINELL HARDNESS

This was the first standard hardness test to be widely accepted. It is rarely used today, having been superceded by other hardness tests, such as the Vickers or Rockwell tests.

The Brinell hardness, H_B, is calculated directly from the geometry of the indentation. It uses a ball indenter, as shown in Figure 2.9. Using a standard 10-mm sphere of steel or tungsten carbide and an applied load of 3000 kg of

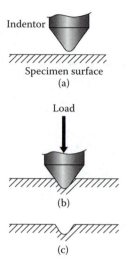

Indentor

Specimen surface
(a)

Load

(b)

(c)

FIGURE 2.8 Example of the procedure for an indentation test.

force, the diameter of the indentation is a measure of the hardness of the test material [7]. Thus,

$$H_B = \frac{F}{\pi D t} = \frac{2F}{\pi D \left[D - \sqrt{D^2 - d^2} \right]}, \tag{2.27}$$

where F = load in kilograms, D = diameter of ball indenter in millimeters, d = diameter of indentation in the surface in millimeters, and t = depth of indentation

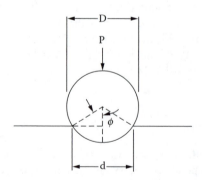

FIGURE 2.9 A spherical indenter produces a circular region on the surface of the material under test. The diameter of the circular indentation is a measure of the hardness of the material.

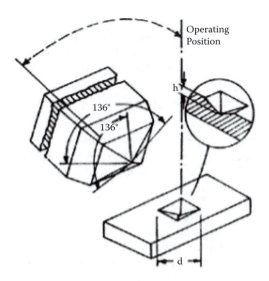

FIGURE 2.10 Diagram of the Vickers Diamond Pyramid Hardness (DPH) indenter.

in surface in millimeters. The typical range of Brinell hardness values is from 5 to 10,000. Notice that the dimensions of Brinell hardness are force per unit area.

2.3.2 VICKERS HARDNESS

This is also known as diamond pyramid hardness. It uses an indenter as shown in Figure 2.10, and the hardness is calculated from [7]

$$H_V = \frac{2F \sin \alpha/2}{d^2} = \frac{1.8544\,F}{d^2}, \qquad (2.28)$$

where F is the load in kilograms, which typically lies in the range 1 to 120 kg for the Vickers test; d is the length in mm of the diagonals of the equiaxed diamond-shaped indentation; and α is the angle between opposite faces of the indenter, which is 136 degrees. The range of values of Vickers hardness is 5 to 1500.

2.3.3 ROCKWELL HARDNESS

The Rockwell hardness is determined simply from the depth of the indentation under specified conditions of applied load. There are several Rockwell hardness scales, and various indenter shapes and loads are used in the different scales, including ball-, diamond-, and brale-shaped indenters and loads of 15 to 150 kg. The depth of indentation is chosen, because measurement of the depth of indentation lends itself easily to simplification and automation of the process, which is one of the attractive features of the Rockwell tests.

Of course, for a particular standard geometry, such as a ball indenter, this depth is an indirect measure of the area of the indentation, so, in principle, the results of the Rockwell test can be traced back to the results of other indentation tests. The Rockwell hardness number is calculated from the equation [7]

$$H_R = M - \frac{t}{0.002},$$
(2.29)

where t is the penetration depth measured in mm. M is the maximum allowed value of the particular scale in use. $M = 100$ for diamond indenters used on the Rockwell scales A, C, and D; whereas $M = 130$ for ball indenters used on the Rockwell scales B, E, M, and R.

2.3.4 KNOOP MICROHARDNESS

The Knoop microhardness test is useful in situations where a small indention is desired or necessary. The term *microhardness* is used for tests that make indentations with loads of value less than 1 kg. The Knoop test uses a load of 300 g on a brale-shaped indenter, as shown in the Figure 2.11, which is similar to the indenter used in the Vickers hardness test. This indenter produces an elongated diamond-shaped indentation with axes of different lengths. The Knoop hardness is then calculated from the equation [7]

$$H_K = \frac{14.2\,F}{2}$$
(2.30)

where F is the load and l is the length of the long diagonal of the indentation. This formula is simply the ratio of the load to the unrecovered (i.e., plastically

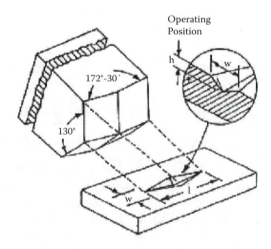

FIGURE 2.11 Details of the Knoop microhardness indenter.

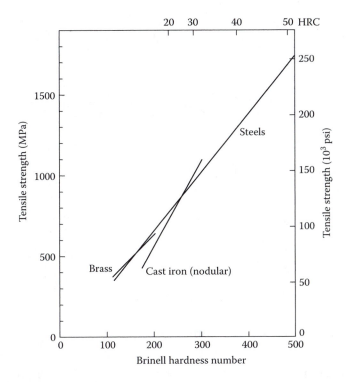

FIGURE 2.12 Relation between hardness and strength.

deformed) projected area of the indentation. The value 14.2 relates the projected area of the indentation to the square of the length of the long diagonal. The range of values of Knoop hardness is from 100 to 1000.

2.3.5 RELATIONSHIP BETWEEN HARDNESS AND OTHER MECHANICAL PROPERTIES

As may be expected, there is a relationship between the hardness, the yield strength, and the ultimate tensile strength of a material. Such a relationship is depicted in Figure 2.12.

EXERCISES: MECHANICS AND MECHANICAL PROPERTIES OF MATERIALS

2.1 The stress–strain curve of AISI 1095 steel is shown in Figure 2.13. The diameter of the test specimen before testing was 9.07 mm, and the minimum diameter after testing in the necked region was 8.18 mm. Determine (1) the elastic modulus, (2) the yield strength for 0.2% plastic strain, (3) ultimate tensile strength, and (4) the ductility from (a) elongation and (b) reduction in area.

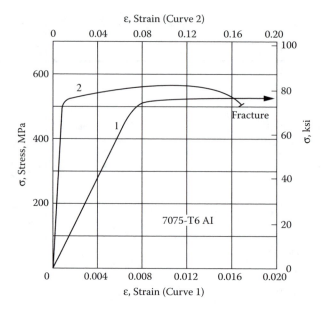

FIGURE 2.13 Stress-strain curve for AISI 1095 steel.

2.2 A specimen of steel with Young's modulus 200 GPa and Poisson's ratio of 0.3 is subjected to a compressive stress of 200 MPa along the z-direction, but is constrained so that it cannot deform in the y-direction, as shown in Figure 2.14. If the material is isotropic, determine (1) the stress in the y-direction, and (2) the strain in the z-direction.

2.3 A certain type of steel has an ultimate tensile strength of 1500 MPa. Estimate its hardness on both the Rockwell C scale and the Brinell scale. A part made of the same steel is impacted by a 10-mm diameter tungsten carbide sphere, which

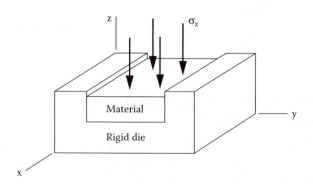

FIGURE 2.14 Diagram showing the shape of the steel part for problem 2.2.

is harder than the steel, with a force of 1000 kg weight (9.8 kN). Calculate the diameter and depth of the resulting indentation in the surface.

REFERENCES

1. Moosbrugger, C., *Atlas of Stress Strain Curves*, 2nd ed., ASM International, Materials Park, OH, 2002, pp. 1–19.
2. Dieter, G.E., *Mechanical Metallurgy*, 3rd ed., McGraw-Hill, New York, 1986.
3. Atkin, R.J. and Fox, N., *An Introduction to the Theory of Elasticity*, Longman, London, 1980.
4. Pilkey, W.D., *Formulas for Stress Strain and Structural Matrices*, John Wiley & Sons, New York, 1994.
5. Chandler, H., *Hardness Testing*, 2nd ed., ASM International, Materials Park, OH, 1999.
6. *Fundamentals of Indentation Hardness Testing*, ASM, Metals Park, OH, 1974.
7. Tabor, D., *The Hardness of Metals*, Clarendon Press, Oxford University Press, Oxford, U.K., 2000.

FURTHER READING

Bowman, K., *Mechanical Behavior of Materials*, John Wiley & Sons, Hoboken, NJ, 2004.

Chalmers, B., *The Structures and Properties of Solids*, Heyden, London, 1982.

Cottrell, A.H., *The Mechanical Properties of Matter*, John Wiley & Sons, New York, 1964.

Davis, J.R. (Ed.), *Metals Handbook: Mechanical, Wear and Corrosion Testing*, ASM, OH, 1998, pp. 1308–1355.

Davis, J.R. (Ed.), *Metals Handbook: Properties of Metals*, ASM, OH, 1998, pp. 114–121.

Eisenstadt, M., *Introduction to Mechanical Properties of Materials*, Macmillan, New York, 1971.

Herzberger, R.W., *Deformation and fracture Mechanics of Engineering Materials*, John Wiley & Sons, New York, 1976.

3 Sound Waves: Acoustic and Ultrasonic Properties of Materials

This chapter discusses the basis for materials evaluation methods that rely on the use of sound waves of different frequencies, including frequencies above the audible range (ultrasound). There are several variants on the theme of the use of ultrasonics in materials evaluation. These include the use of ultrasonics in measurement of thickness, in which the time of flight of an ultrasonic wave front through a material is measured, and, if the velocity is known, then the distance traveled can be calculated. Ultrasonics can be used to determine the elastic modulus of the material, assuming that the material density is known. Ultrasonic resonance can be used, at lower frequencies, to determine elastic properties, speed of sound, or sample dimensions. The reflection of an ultrasonic wave front from an interface or discontinuity can be used to detect the presence of flaws, such as defects, cracks, or other inhomogeneities in materials; and, finally, ultrasonics can be used to produce an image of the location of inhomogeneities in materials.

3.1 VIBRATIONS AND WAVES

The types of waves that can be generated in material come in two forms — transverse waves in which the displacements are perpendicular to the direction of travel of the waves, and longitudinal, in which the displacements are along the direction of travel of the waves, as shown in Figure 3.1.

3.1.1 THE WAVE EQUATION

The amplitude of a wave as a function of time and position can be represented in several equivalent ways, for example, as a sine function, as a cosine function, as an exponential of an imaginary number, or as a second differential equation in which the acceleration is proportional to the displacement. A wave propagating in one dimension inside a material has the general form

$$y(x, t) = A \cdot f(kx + 2\pi v t), \qquad (3.1)$$

where y is the displacement, A is the amplitude of the wave, f is a periodic function, x is the distance along the direction of propagation, t is time, v is the frequency of the wave, and k is the wave vector, the reciprocal of wavelength λ,

Transverse Wave

Longitudinal Wave

FIGURE 3.1 Transverse and longitudinal waves in a solid.

which describes how quickly the displacement varies with distance along the direction of propagation.

In the simplest case, f is a sinusoidal function such as sine or cosine.

$$y(x, t) = A \cdot \cos(kx + 2\pi v t) \tag{3.2}$$

$$\frac{\partial y}{\partial t} = -2\pi v A \cdot \sin(kx + 2\pi v t) \tag{3.3}$$

$$\frac{\partial^2 y}{\partial t^2} = -(2\pi v)^2 A \cdot \cos(kx + 2\pi v t) \tag{3.4}$$

$$\frac{\partial y}{\partial x} = -kA \cdot \sin(kx + 2\pi v t) \tag{3.5}$$

$$\frac{\partial^2 y}{\partial x^2} = -k^2 A \cdot \cos(kx + 2\pi v t) \tag{3.6}$$

and from the above equations it follows that

$$\frac{\partial^2 y}{\partial x^2} = \frac{k^2}{(2\pi v)^2} \frac{\partial^2 y}{\partial t^2}. \tag{3.7}$$

Because $k = 2\pi/\lambda$ and $v\lambda = V$

$$\frac{\partial^2 y}{\partial x^2} = \frac{1}{V^2} \frac{\partial^2 y}{\partial t^2}, \tag{3.8}$$

TABLE 3.1
Velocities and Wavelengths of Ultrasound in Different Materials

	(m.sec⁻¹)	λ (mm) at 1 MHz
Aluminum	6350	6.35
Steel	5850	5.85
Magnesium	5790	5.79
Copper	4660	4.66
Plexiglass	2670	2.67
Polyethylene	1950	1.95
Water	1490	1.49
Oil	1380	1.38
Air	330	0.33

where V is the velocity. This last equation is known as the *wave equation*, although, in fact, it is a special case, which results in a wave with simple harmonic motion.

3.1.2 WAVELENGTH AND FREQUENCY

Although the velocity V of ultrasonic and sonic waves in materials does not change much with frequency, the wavelength λ is inversely proportional to frequency ν,

$$V = \nu\lambda. \tag{3.9}$$

The selection of the wavelength (and thereby, frequency) of waves for studying certain problems in nondestructive evaluation is very important. Typically, the wavelength of the signals should be comparable with the size of the features being studied — for example, cracks or voids in a material. If the wavelength is too long, then the features can be overlooked, or, at best, the perturbation of the signal caused by the feature is not clear.

Examples of wavelengths of longitudinal ultrasonic waves at 1 MHz in different materials are shown in Table 3.1. The velocities of shear waves are typically one half of the velocities of longitudinal waves in the same material.

3.2 RELATIONSHIP BETWEEN MECHANICAL PROPERTIES AND WAVE PROPAGATION

We have seen in Chapter 1 that the elastic modulus is a measure of how a material strains under the action of a stress when below the yield strength or elastic limit. Because a wave is also a response (displacement) of a material to a stress (in this

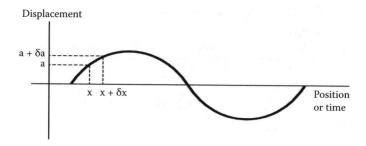

FIGURE 3.2 Depiction of a transverse wave passing through a material along the x-direction. The displacement perpendicular to the direction of travel is a(x).

case, a periodic loading), it is perhaps not surprising to find that the elastic modulus is related to the velocity of wave propagation.

3.2.1 TRANSVERSE WAVES

In the case of transverse waves, the displacement is orthogonal to the direction of propagation of the wave. In solids, these are referred to as shear waves. Consider a transverse wave as shown in Figure 3.2.

The transverse strain e in an elemental section of the material as a transverse wave passes is written as

$$e_{\perp} = \frac{\partial a}{\partial x}. \qquad (3.10)$$

The elastic shear modulus is

$$G = \frac{\sigma_{\perp}}{e_{\perp}}. \qquad (3.11)$$

Differentiation of this equation gives

$$\frac{\partial \sigma}{\partial x} = G\frac{\partial e}{\partial x} = G\frac{\partial^2 a}{\partial x^2}, \qquad (3.12)$$

and stress σ is force/area, therefore $\frac{\partial \sigma}{\partial x}$ is equivalent to mass times acceleration divided by volume

$$\frac{\partial \sigma}{\partial x} = \rho\frac{\partial^2 a}{\partial t^2} \qquad (3.13)$$

$$\frac{\partial^2 a}{\partial t^2} = \frac{G}{\rho}\frac{\partial^2 a}{\partial x^2}. \qquad (3.14)$$

Therefore referring to equation 3.8, the velocity of transverse waves is given by

$$V_S = \sqrt{\frac{G}{\rho}}, \qquad (3.15)$$

so the velocity of a transverse mechanical vibration depends only on the shear modulus and the density of the material. This means that the measurement of wave velocity can be used to determine the elastic modulus assuming that the density is known.

3.2.2 Longitudinal Waves

A similar analysis in the case of a longitudinal wave reveals a similar result. The longitudinal strain, e, is given by the rate of change of displacement with position,

$$e_{//} = \frac{\partial a}{\partial x}, \qquad (3.16)$$

and the longitudinal elastic modulus is given by

$$Y = \frac{\sigma_{//}}{e_{//}} \qquad (3.17)$$

Using the same argument to that used above, the velocity of longitudinal waves is given by

$$V_L = \sqrt{\frac{Y}{\rho}}. \qquad (3.18)$$

Interestingly, it is therefore not necessary to perform a tensile test to measure the elastic moduli of a material. Both shear and longitudinal moduli can be determined depending on the type of wave used. In crystalline materials, different values of acoustic velocity are seen and, hence, different values of the elastic moduli Y and G are found along different crystallographic directions.

In practical terms, for most nondestructive evaluation purposes, it is assumed that the material is ideally a homogeneous continuum. In reality it is not. Materials are discrete assemblies of atoms, and when materials have been subjected to service degradation, they can have cracks. However, for low frequencies (in which the wavelengths are long compared to interatomic spacings), the continuum approximation is adequate. As it happens, the same relationship between wave velocity and

TABLE 3.2
Velocity of Longitudinal and Shear
Waves in Different Materials

Material	V_L (m.sec^{-1})	V_S (m.sec^{-1})
Aluminum	6400	3000
Copper	4700	2300
Iron	5900	3200
Nickel	6000	3000
Pyrex glass	5600	3300
Flint glass	4000	2400
Crown glass	5100	2800
Polystyrene	2400	1100

elastic modulus is obtained for low frequencies whether one uses an atomistic or a continuum approach.

Table 3.2 shows values of sound velocity in various materials. Note that there is not much variation from material to material, with the velocities all being in the range of a few thousand meters per second, with longitudinal wave velocities being typically twice that of transverse (shear) waves.

WORKED EXAMPLE

An ultrasonic pulse passing through a 5 cm long copper rod returns to the transducer after 21 μsec. Is there a flaw in the rod if the wave is longitudinal? Is there a flaw in the rod if the wave is transverse? Assume Young's modulus is 2×10^{11} Pa, Poisson's ratio is 0.3, and the density is 9000 kg.m^{-3}.

Solution

For longitudinal waves

$$v = \sqrt{\frac{Y}{\rho}} = \sqrt{\frac{2 \times 10^{11}}{9000}} \tag{3.19}$$

$$v = 4.71 \times 10^3 \text{ m} \cdot \text{sec}^{-1}, \tag{3.20}$$

and the distance traveled by the wave is

$$d = vt = 0.099 \text{ m}, \tag{3.21}$$

which is, to within an error of 1%, twice the length of the rod. Therefore, if it is a longitudinal wave, then the echo that is being detected is simply the echo from the end of the rod, and there is no flaw.

For shear waves

$$v = \sqrt{\frac{G}{\rho}} = \sqrt{\frac{Y}{2(1+v)\rho}} = \sqrt{\frac{2 \times 10^{11}}{(2.6)(9000)}} \tag{3.22}$$

$$v = 2.92 \times 10^3 \text{ m} \cdot \text{sec}^1, \tag{3.23}$$

and the distance traveled by the wave is

$$d = vt = 0.061 \text{ m}. \tag{3.24}$$

Therefore, if it is a shear wave, there is a flaw at a depth of 0.0305 m below the surface.

3.2.3 CHANGES IN MECHANICAL PROPERTIES

The elastic moduli of materials, and hence the ultrasonic velocity, can change as a result of stress or other types of degradation in materials. These changes are small (usually less than 1%), but can be detected fairly easily with ultrasonic equipment via velocity changes, which can be detected down to parts per million.

For example, the elastic modulus changes with stress/strain according to the following equation

$$Y = Y_o + \frac{dY}{de} de. \tag{3.25}$$

The term dY/de is the third-order elastic modulus, which means that it is the third derivative of the elastic energy with respect to strain. Higher-order elastic moduli can also be defined in terms of the higher order derivatives of elastic energy with respect to stress, although use of these higher-order elastic moduli is extremely rare.

In principle, the third-order elastic moduli could be used for stress evaluation, although it seems that this has not actually been employed in practice for such an application. However, elastic modulus determination can still be useful for other types of material evaluation purposes.

There is also a relationship between amplitude of ultrasound scattered from defects in a material and the ductility of the material [1]. This means that the ductility can be determined without the need to stress the material to failure.

3.3 LAUNCHING WAVES IN MATERIALS

Launching a mechanical vibration into a material and observing the results is the most widespread method employed in materials evaluation [2,3]. The propagation of the sound waves allows us to:

1. Look for unexpected echoes, which may be reflections of sound wave fronts from flaws (internal surfaces).
2. Look for unexpected echoes from external surfaces, which may indicate changes in thickness of a material or part.
3. Measure changes in velocity of acoustic or ultrasonic waves, which can indicate structural changes, stress, etc.

Different ultrasonic wavelengths/frequencies are used to investigate different types of problems in materials. For example, assuming a typical ultrasonic velocity of 5000 m·sec^{-1}, the following wavelengths/frequencies are used to determine specific issues:

25 mm – (200 KHz–MHz) is used for studying coarse grain castings;
12.5–1 mm (400 kHz–5 MHz) is used for studying fine grain castings;
5 mm – (1–5 MHz) is used for rolled products, metal sheets, plates, and bars;
5–0.5 mm (1–10 MHz) is used for forgings;
5–2 mm (1–2.5 MHz) is used for welds;
5–0.5 mm (1–10 MHz) is used for cracks.

3.3.1 TRANSDUCERS

To get mechanical vibrations in materials, transducers are used. These are devices that can be used to launch waves in materials by converting electrical impulses into mechanical impulses [4]. Normally, these are piezoelectric materials, which respond to an applied voltage at a given frequency by straining at the same frequency. Such materials include quartz, lead zirconate titanate (PZT), and lithium niobate. However, other materials can also be used for this purpose, including magnetostrictive materials, such as terfenol-D, which undergoes large strains when subjected to a magnetic field.

3.3.2 MODES OF INSPECTION: PULSE-ECHO AND PITCH-CATCH

There are several ways to inspect a material for flaws using ultrasound, but the most popular is still the method in which the same transducer is used to launch the ultrasonic pulse into the material and to detect the echo. This is known as the *pulse-echo mode*. This situation is depicted in Figure 3.3.

An alternative method is to use two transducers—one to launch the pulse and the other to detect the echo. This is known as *pitch-catch mode* and is depicted in Figure 3.4.

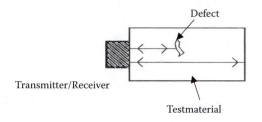

Defect

Transmitter/Receiver

Testmaterial

(a)

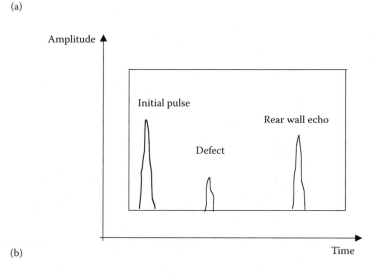

Amplitude

Initial pulse

Rear wall echo

Defect

(b)

Time

FIGURE 3.3 Single-ended transducer method operating in "pulse-echo" mode.

The orientation of the transducers does not need to be at the ends of the sample, as shown in Figure 3.4. It is also possible to launch and detect ultrasonic wave pulses with two transducers on the same surface, as shown in Figure 3.5 with "angle beams."

In all of the above cases, the objective is to find any echoes that are reflected from flaws in the material. These flaws can be thought of simply as internal

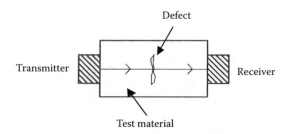

Defect

Transmitter

Receiver

Test material

FIGURE 3.4 Double-ended transducer method operating in "pitch-catch" mode.

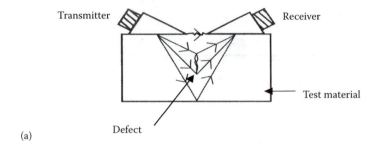

(a)

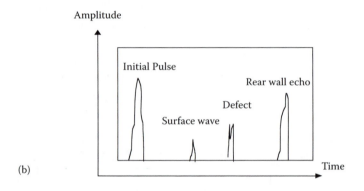

(b)

FIGURE 3.5 Generation and detection of ultrasonic vibrations using angle beams.

surfaces from which the wave front is reflected. The result is that there are echoes arriving at the transducer at times when there is not supposed to be such a signal; therefore indicating that reflection of the ultrasound is occurring from locations at which there should be no reflections. This is normally indicative of flaws.

A pulse-echo wavetrain is detected and displayed on an oscilloscope. Not all echoes are necessarily indicative of problems. For example, reflection from the back surface (back wall) of a sample is often expected for waves propagating normal to the back wall, and this is shown in Figure 3.6. However, the presence of an echo in between the initial pulse and the reflection from the back wall is usually indicative of a problem, and, most likely, indicates the presence of a flaw between the front surface and back wall.

3.3.3 TIME OF FLIGHT: THICKNESS DETERMINATION

If the velocity of sound in a material is known, then a pulse-echo method can be used to determine the thickness of the material [5]. Consider the following situation in which there has been some loss of material on the far side of a test specimen. In cases where the far side is not accessible or visible, the use of ultrasound to inspect the thickness of the material may be the best solution. The wave with velocity v is launched perpendicular to the surface from the nearside, and the

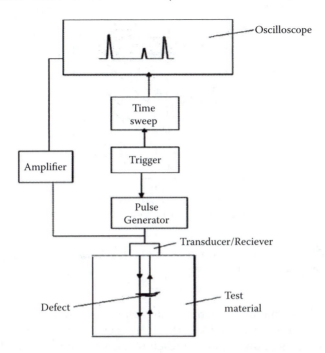

FIGURE 3.6 (a) Block diagram of the electronics used to launch and detect ultrasonic waves in a test specimen. (b) Pulse-echo diagram of signal intensity vs. time, showing the back wall reflection and the presence of an intermediate echo that could be indicative of a flaw.

time, t, for the echo to return is measured. The thickness, d, is then determined at different locations over the surface.

Therefore, by merely looking at the time taken for the ultrasonic pulse to return from the back wall, the thickness can be determined by (Figure 3.7):

$$2d_1 = vt_1 \tag{3.26}$$

$$2d_2 = vt_2. \tag{3.27}$$

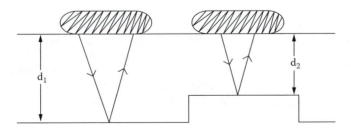

FIGURE 3.7 Sample showing loss of material on the lower side of the test specimen.

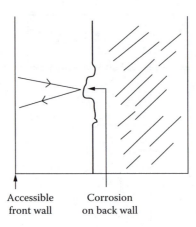

Accessible   Corrosion
front wall   on back wall

FIGURE 3.8 Corrosion (left) on an inaccessible back wall can be monitored by ultrasonic inspection from the accessible front wall.

This method of thickness determination is particularly useful in situations where there is corrosion on inaccessible surfaces, as shown in Figure 3.8, so that the thickness is changing over time as a result of a degradation process. The use of periodic ultrasonic inspection can be used to determine when the loss of material has reached a critical stage.

3.3.4 ATTENUATION

The passage of waves in a material is subject to attenuation with distance and time, resulting from such factors as scattering of the wave energy, absorption of some of the energy by the material, and divergence [6]. If a wave is attenuated exponentially inside a material — for example, if there is a plane wave front, which is not therefore subject to divergence — then the wave equation can be modified to take the attenuation into account as follows,

$$y(x, t) = A_o \sin(kx + 2\pi v t) \exp(-\alpha x - \beta t), \tag{3.28}$$

where the exponential term ensures that the signal decays with both time and distance. This equation only applies in special cases in which the material is homogeneous, and in which the sound beam is not diverging. In many instances, the concern is only one of decay of amplitude with distance, but, at a constant velocity, the exponential decay of amplitude with distance also results in an exponential decay of amplitude with time. This can be shown on an oscilloscope.

If there is no beam divergence, and if the only mechanisms for the decrease in intensity with distance are scattering and absorption, the decay with distance of the sound intensity can be expressed in a particularly simple exponential form,

$$I(x) = I(0) \exp(-\mu x), \tag{3.29}$$

where $I(x)$ is the intensity of the sound (the energy per unit area per unit time in $J.m^{-2}.sec^{-1}$, which is proportional to the amplitude squared in the wave equations above) at a distance x into the material, $I(0)$ is the intensity at $x = 0$, and μ is the attenuation coefficient, which is a materials property.

Both absorption and scattering can contribute to μ. Absorption can be affected by damping of the acoustic vibrations in the sample, such as that caused by dislocations and other defects. Scattering can be affected by grain size in the material. For this reason, it is incorrect to call this the "absorption" coefficient. The correct term is the *attenuation* coefficient.

This equation is often used to describe the decrease in intensity of acoustic or ultrasonic intensity with distance, but it is also worth remarking that the equation is not generally applicable. Note that divergence of the wave front is not taken into account in this equation. So this equation really only applies in the case of a collimated (meaning undivergent) beam.

In practice, calibration standards, which usually comprise a series of flat-bottomed holes in a specimen of a particular material, are used to determine attenuation of ultrasound in the same material [7].

WORKED EXAMPLE

An ultrasonic pulse in a plate returns to the transducer in 20×10^{-6} sec. If the velocity of sound is 4.67×10^3 m · sec^{-1}, and the intensity of the echo is 50% of the original signal, what is the attenuation coefficient?

Solution

The distance traveled, x, is given by

$$x = vt = (4.67 \times 10^3).(20 \times 10^{-6}) \tag{3.30}$$
$$= 0.0934 \text{ m}$$

$$\mu = -\frac{1}{2x} \log_e \left(\frac{I}{I_o} \right) \tag{3.31}$$

$$\mu = 3.71 \text{ m}^{-1} \tag{3.32}$$

3.3.5 NOTATION FOR ATTENUATION AND AMPLIFICATION OF SIGNALS

The acoustic power, measured in term of watts per square meter, is proportional to the square of the signal amplitude. The convention for measuring changes in signal strength is the so-called decibel notation. If the intensities of two signals are written as I_1 and I_2, then the ratio of these can be expressed as

$$\frac{I_1}{I_2} = 10^{N_B} \qquad (3.33)$$

so that

$$N_B = \log_{10}\left(\frac{I_1}{I_2}\right), \qquad (3.34)$$

where N_B is the number of bels. The number of decibels N_{DB} is then simply ten times this number:

$$N_{DB} = 10 \log_{10}\left(\frac{I_1}{I_2}\right). \qquad (3.35)$$

3.3.6 ACOUSTIC EMISSION

Sound can also be generated by crack growth in materials without the presence of transducers. This is known as *acoustic emission* and is a viable form of nondestructive testing, which allows the growth of cracks to be monitored from the sounds that they emit while the material is undergoing such changes [8].

3.3.7 LASER GENERATION OF ULTRASOUND

Another method for generating ultrasound in materials is through the use of laser ablation and thermoelastic expansion [9]. This method has the advantage that no physical contact with the test sample is needed. On the other hand, the signal levels produced tend to be low compared with other methods of generation of ultrasound.

EXERCISES: SOUND — SONICS AND ULTRASONICS

3.1 A sample of steel with a flat surface is examined using shear waves with a 50°-angle probe (meaning the beam is at 50° to the normal inside the test material). The shear wave velocity in the steel is 3,000 m · sec^{-1}. A defect echo is detected at 46.6 μsec. What is the depth of the defect? What is the "stand-off" distance (the component of distance of the flaw from the transducer parallel to the surface)?

3.2 A 5-MHz longitudinal (compression) wave probe is used to examine a piece of steel. What is the wavelength of the signal? If the same probe is used with a piece of lead, what is the wavelength. Give approximate wavelengths for 5-MHz shear waves in these materials (by making an estimate of the shear wave velocity). The velocity of compressional sound waves in steel is 5960 m · sec^{-1}, and in lead is 1960 m · sec^{-1}.

3.3 Ultrasonic waves of frequency 10 MHz are generated in a part of unknown thickness made of a material with elastic modulus 2×10^{11} Pa and density 8000 kg m^{-3}. An acoustic signal is detected from the back wall reflection after 4 μsec with an intensity of 80% of the original signal. Calculate the wavelength of the ultrasonic signal, the thickness of the part, and the attenuation coefficient for the ultrasound, assuming negligible divergence of the wave front.

REFERENCES

1. Tittmann, B.R., Ultrasonic measurements for the prediction of strength, *NDT Int* 11, 17, 1978.
2. Wells, L.H., Basic ultrasonics: 1 — types of wave, reflection, and the decibel, *Nondestr Test Eval Int* 1, 233, 1968.
3. Wells, L.H., Basic ultrasonics: 2 — the use of compression (longitudinal) waves in ultrasonic testing, *Nondestr Test Eval Int* 1, 291, 1968.
4. Smith, A.L., Ultrasonic fundamentals, *Mater Eval* 36, 37, 1978.
5. Cartwright, D.L., Ultrasonic thickness measurements of weathering steel, *Mater Eval* 53, 452, 1995.
6. Burkle, W.S. and Isom, V.H., Use of DAC curves and transfer lines to maintain ultrasonic test references level sensitivity, *Mater Eval* 41, 624, May 1983.
7. Beck, K.H., Limitations to the use of reference blocks for periodic and preinspection calibration of ultrasonic instruments and systems, *Mater Eval* 57, 323, 1999.
8. Richter, J.R., Acoustic emission testing, *Mater Eval* 57, 492, 1999.
9. Hopka, S.N. and Ume, I.C., Laser ultrasonics: simultaneous generation by means of thermoelastic expansion and material ablation, *J NDE* 18, 91, 1999.

FURTHER READING

Auld, B.A., *Acoustic Fields and Waves in Solids*, Vol. I and II, 2nd ed., Krieger Publishing Company, February 1990.

Bradfield, G., Elasticity measurements using a critical angle reflection technique, *NDT Int* 1, 370, 1968.

http://www.ndt-ed.org/EducationResources/CommunityCollege/Ultrasonics/cc ut_index. htm.

Krautkramer, J. and Krautkramer, H., *Ultrasonic Testing of Materials*, 4th ed., Springer-Verlag, Berlin, 1990.

Silk, M.G., *Ultrasonic Transducers for Nondestructive Testing*, Adam Hilger, Bristol, 1984.

Wells, L.H., Basic ultrasonics, *NDT Int* 1, 233, 1968 and 1, 291, 1968.

4 Thermal Properties of Materials

In this chapter, we discuss the effects of heat on materials, including thermal properties and changes in the structure of materials that arise because of exposure to heat. These characteristics of materials also form the basis of materials evaluation methods that depend on the use of thermal properties of materials and thermal measurement techniques. Thermal inspection measurements are usually made using infrared detection equipment, for which a range of techniques and specialized terminology have been developed, including such methods as ultrasonic generation of heat in test materials, a method known as *ultrasonic thermography*.

4.1 THERMAL EFFECTS IN MATERIALS

We first consider what effects heat has on materials. These can be classified as reversible and irreversible effects. Reversible effects are those in which the amount of heat transferred to the material is sufficiently small that no structural changes occur in the material. Examples of thermal effects that can be reversible are thermal conduction, thermal expansion, changes to elastic moduli, and changes to ductility, such as the ductile-to-brittle transition caused by changes in temperature.

Irreversible thermal effects are those for which the amount of heat transferred to the material are large enough to cause substantial changes. These include thermally induced changes in microstructure (including grain growth), thermal embrittlement of the material, stress relief (annealing of dislocations), recrystallization, reduction of hardness, and thermal shock, the last of which can actually cause failure.

Thermal inspection measurements are usually made using infrared detection equipment, for which a range of techniques and specialized terminology have been developed [1].

The Seebeck effect is often used to measure temperatures in "contact mode" [2]. Detection of infrared radiation is the main alternative method. Both of these can be used to provide an image of the variation of temperature across the surface of a component [3].

4.1.1 THERMAL CAPACITY AND TEMPERATURE CHANGE

The temperature, T, of a material is simply a measure of the energy of vibration, E, of the atoms, according to the relation $E = 3k_BT$, where k_B is Boltzmann's

TABLE 4.1
Typical Values of Heat Capacity

Material	$C_m(\text{J.kg}^{-1} \cdot \text{C}^{-1})$
Aluminum	900
Iron	450
Alumina (Al_2O_3)	775
Polyethylene	1850
Polystyrene	1170
Tungsten	138

constant. The rate of increase in temperature with the amount of energy supplied is the heat capacity, which is normally measured in Joules per degree Celsius, or sometimes in calories per degree Celsius.

$$C = \frac{Thermal\ energy\ supplied}{Change\ in\ temperature} \tag{4.1}$$

$$C = \frac{dQ}{dT}. \tag{4.2}$$

Often, the specific heat is quoted, which can be expressed either as the heat capacity per unit volume or as the heat capacity per unit mass C_m (Table 4.1):

$$C_m = \frac{1}{m}\frac{dQ}{dT}. \tag{4.3}$$

The equilibrium temperature of materials may vary when subjected to the same amount of heat due to differences in heat capacity. The transient temperature can vary because of differences in thermal conduction, due either to intrinsic differences in component materials, or to the localized influence of defects on thermal conduction. Any material body emits electromagnetic radiation with the wavelength characteristic of its temperature and surface emissivity. Variation of temperature across the surface of a component can be used to generate a thermal image.

4.1.2 THERMAL CONDUCTION

When a temperature gradient is applied to a material, heat will be conducted through the material down the temperature gradient. Under steady-state conditions, the process is governed by a well-known equation for the steady-state phase of diffusion known as Fick's First Law. This law relates the rate of flow of a quantity Q per unit area per unit time $(1/A)(dQ/dt)$ to the concentration gradient of that quantity $(1/V)$ (dQ/dx) driving the diffusion, according to the equation

$$\frac{1}{A}\frac{dQ}{dt} = -D\frac{1}{V}\frac{dQ}{dx}, \tag{4.4}$$

where the coefficient D is the diffusivity which has units of $m^2 \cdot sec^{-1}$. This steady-state diffusion equation can be applied equally successfully to discrete quantities, such as atoms or electrons, or to continuous quantities, such as heat energy. In the case of heat flow, the diffusion equation can be expressed as

$$\frac{1}{A}\frac{dQ}{dt} = -D\frac{1}{V}\frac{dQ}{dx} = -DC_v\frac{dT}{dx} = -K\frac{dT}{dx}, \tag{4.5}$$

where dQ/dt is the rate of flow of heat, A is the cross-sectional area, dT/dx is the thermal gradient, C_v is the specific heat capacity per unit volume, and K is the thermal conductivity. From this, it is seen that the thermal conductivity of a material is related to the diffusivity via the specific heat capacity.

Heat capacity and thermal conductivity are clearly related, but thermal conductivity is more easily adapted to the development of thermal nondestructive evaluation of materials. Transient thermal effects are used in the technique of thermography in which a material is heated, and then the temperature profile is studied over time as it cools back to equilibrium. Regions of slow cooling can be indicative of low thermal conduction in the vicinity, which are often the result of flaws in the material that interrupt the heat flow.

Typical values of the thermal conductivity K are shown in Table 4.2.

Heat transfer in a material occurs through two mechanisms: atomic vibrations (known as *phonons*) and electron transport. This means that the thermal

TABLE 4.2
Thermal Conductivities of Various Materials

Material	K (W · m^{-1}K^{-1})
Al	237
Cu	398
Fe	80
Ag	428
Alumina	39
Si	83
Glass (SiO$_2$)	2
Ceramic pottery	0.1

conductivity of metals is higher than nonmetals, because of the existence of both mechanisms operating simultaneously; whereas, in insulators, the contribution of electron transport to thermal conductivity is negligible. However, thermal conduction by atomic vibrations is always present and, therefore, the difference in thermal conductivity between the most-conducting and the least-conducting materials is not as great as the difference in electrical conductivity of these materials.

The conduction of heat though a material is interrupted if the material contains defects, particularly delaminations or layers in which the material is not continuous. This gives us a basis for materials evaluation methods in which the presence of poorly conducting volumes (flaws) can be detected through changes in thermal conductivity from place to place in a material [4], or through differences in temperature [5].

WORKED EXAMPLE

The walls of a container are 8 mm thick. Inside the vessel, the temperature is 70°C. Outside the vessel it is 69.1°C. If the heat flux is 1395 cal·m^{-2}·sec^{-1} determine the thermal conductivity of the wall material.

Solution

Starting from the equation for the heat flux,

$$dQ = KA \frac{\Delta T}{x} dt \qquad (4.6)$$

$$K = \frac{x}{A\Delta T} \frac{dQ}{dt} = \frac{0.008}{0.9} \cdot 1,395 \qquad (4.7)$$

$$K = 12.4 \ cal \ m^{-1} sec^{-1} C^{-1} = 52 \ J m^{-1} sec^{-1} C^{-1}. \qquad (4.8)$$

4.1.3 THERMAL EXPANSION

When materials are heated, their dimensions can change. In the case of metals, these materials usually experience an increase in length as the temperature increases. This is known as *thermal expansion*. This property of a material is normally represented by the *thermal expansion coefficient* α, which is defined as the fractional change in length per unit temperature change,

$$\alpha = \frac{\Delta}{\Delta T} \cdot \frac{1}{\Delta T}. \qquad (4.9)$$

TABLE 4.3
Thermal Expansion Coefficients for Various Materials

Material	$\alpha(C^{-1})$
Aluminum	24×10^{-6}
Copper	17×10^{-6}
Iron	12×10^{-6}
Invar	0.7×10^{-6}
Alumina Al_2O_3	7.6×10^{-6}
Soda lime glass	9.0×10^{-6}
Polytetrafluoroethylene	180×10^{-6}
Polyethylene	150×10^{-6}

4.1.4 STRESS DUE TO THERMAL EXPANSION

The stress σ induced by thermal expansion α can be calculated from the strain $e = \alpha\Delta T$, and the elastic modulus Y of the material:

$$\sigma = Y\frac{\Delta}{} = Y\alpha\Delta T. \tag{4.10}$$

Therefore, for a given change in temperature, the stress produced is dependent on both the elastic modulus Y and the thermal expansion coefficient α. Typical values of the thermal expansion coefficient are listed in Table 4.3.

4.2 TEMPERATURE DEPENDENCE OF MATERIALS PROPERTIES

Mechanical properties of materials (e.g., fracture toughness and ductility) are dependent on temperature. These can vary nonlinearly with temperature, as shown Figure 4.1.

The low-temperature regime shows *low toughness*, meaning that the material is brittle in this temperature range. Then, at the transition temperature, the toughness increases rapidly with increasing temperature to reach a ductile regime at higher temperatures. The variation of fracture toughness with temperature is reversible, provided that no structural changes have occurred as a result of the temperature increase. So, in most cases, the curve is retraced as the temperature is reduced, but if the material is exposed to elevated temperatures, or to high levels of ionizing radiation over an extended period of time, there can be irreversible structural changes, which will affect the shape of the curve and can alter the transition temperature.

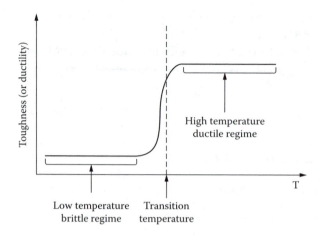

FIGURE 4.1 Variation of fracture toughness (or ductility) with temperature.

4.2.1 DUCTILE-TO-BRITTLE TRANSITION

The ductile-to-brittle transition temperature shown in Figure 4.2 is microstructure-dependent, so that any factor that causes a change in the microstructure of a material can also likely affect the transition temperature. A well-known example of this is the effect of radiation damage on the ductile-to-brittle transition temperature. Exposure over a period of years to high fluence of neutrons in a nuclear reactor, for example, can cause deterioration in microstructure, resulting in a reduction in the transition temperature.

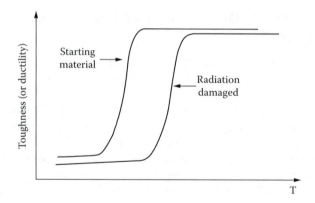

FIGURE 4.2 Dependence on radiation exposure of the variation of fracture toughness with temperature.

This means that a material operating under its normal temperature and pressure conditions may initially be ductile, but if the radiation damage is sufficiently great, the transition temperature may eventually be above the operating temperature, so that the material may fail under operating conditions that it had previously been able to tolerate. Therefore, detecting the shift in the ductile-to-brittle transition temperature can be very important in the effort to avoid catastrophic failure (Figure 4.2).

4.3 EFFECTS OF HEAT ON STRUCTURE OF MATERIALS

At this point, it is relevant to look at how the structure of a material changes with exposure to elevated temperatures. If the temperature is raised to 100°C, 300°C, 500°C, or 700°C, the effects on the microstructure of the material depend as much on the time of exposure as on the temperature, and so both of these factors have to be taken into account.

For convenience, the changes are usually classified into three groups: recovery, recrystallization, and grain growth. The boundaries between these effects are not completely well-defined, because, as the temperature is increased, one process will gradually transition into another, but for the purposes of discussion, the distinction is useful.

4.3.1 RECOVERY, RECRYSTALLIZATION, AND GRAIN GROWTH

Recovery means the relief of stress as a result of the annealing of dislocations. *Recrystallization* means the nucleation of new strain-free (dislocation-free) grains (crystallites) at elevated temperatures. *Grain growth* means the growth of the new strain-free grains that, if the process continues, will ultimately form a large-grained specimen from one that may originally have had a number of smaller grains.

These processes are depicted in the Figure 4.3, in which the recovery process is seen to be occurring at temperatures up to about 200°C, the recrystallization is occurring mostly between 200–500°C, and the grain growth is occurring above 500°C. Simultaneously, there are changes in the mechanical properties of the material, whereby the ductility is improved as a result of annealing and the tensile strength is reduced. Note that this figure does not include the effects of time, which means the figure depicts the progress of these processes over similar time periods. It is also true that the recrystallization and grain growth can reach the same state of development at lower temperatures, if the time allowed for the process is longer.

The process of recrystallization is one in which the atoms are rearranged and, therefore, the process must be dependent on the strength of the interatomic bonds. Because the melting temperature is also a measure of the interatomic bond strength,

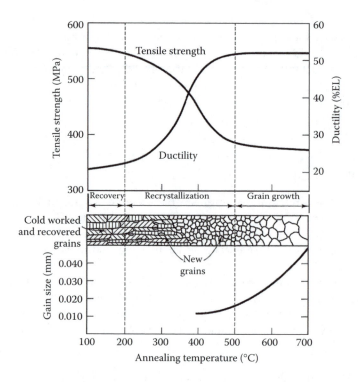

FIGURE 4.3 The various effects of exposure to elevated temperature on materials — recovery, recrystallization, and grain growth.

there is a direct correlation between the so-called recrystallization temperature and the melting temperature. This is depicted in Figure 4.4.

4.3.2 EFFECTS OF ANNEALING ON HARDNESS AND DUCTILITY

Because heat treatment of materials affects the microstructure, it also affects the mechanical hardness and ductility. In this respect, it is often meaningless to talk about the hardness of a type of steel, for example, without discussing the heat treatment. Table 4.4 shows the hardness values for different compositions of steel subjected to a variety of heat treatments.

The microstructure, and therefore the hardness, are also affected by the rate of change of temperature, as can be seen in Table 4.4 for several types of steels. Rod-shaped samples of materials were heated until they reached the high-temperature austenitic phase. Then, the end of the rod was quenched in oil at room temperature known as a Jominy end quench. This means that the rate of cooling was greatest at zero distance from the end, and that the cooling rate decreased as the distance

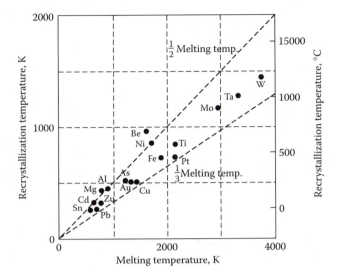

FIGURE 4.4 Variation of recrystallization with melting temperature for various materials.

from the end, increased. The resulting hardness values as a function of distance reflect these differences in the cooling rate, with the highest hardness values at the quenched end, and progressively lower values of hardness at distances along the rod further from that end.

TABLE 4.4
Rockwell C Hardness and Ductility of 2.5-cm Diameter Steel Rods

Alloy Designation/ Quenching Medium	As -Quenched Hardness (HRC)	Tempered at 540°C (1000°F) Hardness (HRC)	Tempered at 540°C (1000°F) Ductility (%EL)	Tempered at 595°C (1100°F) Hardness (HRC)	Tempered at 595°C (1100°F) Ductility (%EL)	Tempered at 650°C (1200°F) Hardness (HRC)	Tempered at 650°C (1200°F) Ductility (%EL)
1040/oil	23	(12.5)[a]	26.5	(10)[a]	28.2	(5.5)[a]	30.0
1040/water	50	(17.5)[a]	23.2	(15)[a]	26.0	(12.5)[a]	27.7
4130/water	51	31	18.5	26.5	21.2	—	—
4140/oil	55	33	15.5	30	19.5	27.5	21.0
4150/oil	62	**38**	**14.0**	35.5	15.7	30	18.7
4340/oil	57	**38**	**14.2**	35.5	16.5	29	20.0
6150/oil	60	**38**	**14.5**	33	16.0	31	18.7

[a] These hardness values are only approximate because they are less than 20 HRC.

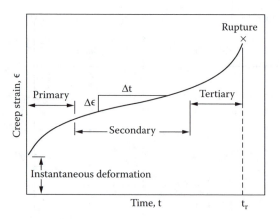

FIGURE 4.5 Rate of creep deformation increase with stress and temperature.

4.3.3 CREEP

Creep damage is a type of failure that occurs when a material is subjected simultaneously to high levels of temperature and stress. In this type of failure, the material undergoes a slow flow over time, and this results in a failure that occurs at stress levels below the ultimate tensile strength. For a given stress level, the failure occurs sooner at higher temperatures. The variation of strain e with time during creep failure is shown in Figure 4.5. Quantitative models of the creep damage process exist and these provide mathematical equations governing the progress of creep.

Creep testing is difficult to carry out in practice, because the time scales that lead to failures are often very long (of the order of tens of years), so that in most cases, the only creep tests that can be performed in a reasonable time are accelerated tests, which are usually carried out at a higher temperature than is encountered in practice, or less often, at a higher stress than is encountered in practice.

The equation that is usually used to describe the variation of creep strain e with time t in the steady-state (secondary) regime is

$$\frac{d\varepsilon}{dt} = K\sigma^n \exp\left(-\frac{Q}{RT}\right), \tag{4.11}$$

where K and n are material constants, Q is the activation energy per mole for creep, T is the thermodynamic temperature, and $R = 8.314$ J·mol^{-1}·K^{-1} is the energy per mole per degree, where a mole is 6.02×10^{23} atoms.

EXERCISES: EFFECTS OF HEAT: THERMAL PROPERTIES

4.1 A piece of aluminum was used to temporarily plug a hole in a piece of steel pipe, giving a perfect fit at 15 °C. The temperature then rose to 60 °C. The thermal expansion of the aluminum is 26×10^{-6} °C^{-1} and its elastic modulus is 0.7×10^{11} Pa; the thermal expansion of the steel is 12×10^{-6} °C^{-1} and its elastic modulus 2.0×10^{11} Pa. Calculate the thermal stress at the interface between the two materials.

4.2 The walls of a steam pipe in a power station are 8 mm thick and are made of steel with thermal conductivity $50 \ W \cdot m^{-1} \cdot °C^{-1}$. The pipe is also protected by insulating lagging of thickness 20 mm and thermal conductivity $4 \ W \cdot m^{-1} \cdot °C^{-1}$. The temperature of the outer surface of the insulating lagging is 40 °C, and the heat flux through the outer surface is measured to be 15,000 W.m^{-2}. Determine the temperature of the interface between the pipe and the insulating lagging, as well as the temperature of the steam in the pipe.

4.3 A 2-kg part made of metal with a ductile-to-brittle transition temperature of 0 °C from an upper shelf energy of 200 J to a lower shelf energy of 10 J is exposed to unexpected cooling which results in a loss of 5000 cal of heat. If the initial temperature of the part was 20 °C and its specific heat capacity is 500 J·kg^{-1}·C^{-1}, determine whether the part is ductile or brittle after the cooling, and decide whether this could be a cause for concern in the continued operation of the part if it operates under load.

REFERENCES

1. C. Hockings, Infrared equipment terminology, *Mater Eval* 55, 955, September 1997.
2. Morgner, W., Introduction to thermoelectric nondestructive testing, *Mater Eval* 49, 1081, 1991.
3. Newitt, J., Application specific thermal imaging, *Mater Eval* 45, 500, May 1987.
4. Bouvier, C., Investigating variables in thermographic composite inspections, *Mater Eval* 53, 544, 1995.
5. Thomas, R.L., Favro, L.D., Grice, K.L., Inglehart, L.J., and Lin, M.J., Thermal wave imaging for NDE of metals, in *NDE of Microstructure for Process Control,* Wadley, H.N.G., Ed., ASM, OH, 1985, p. 119.

FURTHER READING

Wright, H.C., *Infrared Techniques*, Clarendon Press, Oxford, 1973.

5 Electrical and Magnetic Properties of Materials

The electrical and magnetic effects in materials are closely related because both are dependent on the electrons. Many of these properties are structure insensitive, so that they are not useful for the purposes of nondestructive evaluation of structure of materials. We must, therefore, identify the magnetic and electrical properties that may be of interest for our purposes. We first ask what happens when a material is subjected to an electric field, and then, what happens when the same material is subjected to a magnetic field. From these responses, it is possible to identify measurements that can be used for nondestructive evaluation.

5.1 ELECTRICAL INSULATORS

When an insulator (dielectric) is subjected to an electric field, this results in electric polarization, meaning the buildup of electrical charges on the surfaces along the direction of the field.

5.1.1 POLARIZATION

Polarization is a measure of how much a material becomes charged when subjected to a particular field strength. Consider a slab of material as shown in Figure 5.1.

If the total charge on the end face is q coulombs, and the charge separation distance is d meters over a volume V, the polarization P is

$$P = \frac{q \cdot d}{V} = \frac{q}{A}.$$ (5.1)

The units of polarization are coulombs per square meter.

5.1.2 RELATION BETWEEN POLARIZATION AND FIELD

In many materials, the polarization P depends linearly on the electric field strength E, according to the equation:

$$P = \varepsilon_o \chi_e E,$$ (5.2)

where ε_o is the dielectric coefficient, and χ_e is the electric susceptibility. There is a general constitutive equation between electric field E, polarization P, and the

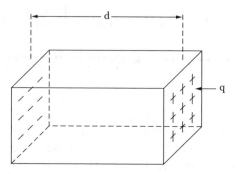

FIGURE 5.1 How a material becomes electrically charged.

total electric flux density D, in which the electric flux density is a vector sum of the polarization and the electric field according to the equation

$$D = \varepsilon_o E + P. \tag{5.3}$$

In this case, the permittivity of space ε_o has a finite value. In other words, when an electric field E is applied in free space, an electric flux density D is generated; it is not necessary to have a material present to get this effect. However, without a material, there is no polarization P.

5.1.3 SURFACE CHARGE

It follows from the above that the electric charge on a surface q is given by

$$q = \chi_e \varepsilon_o E A. \tag{5.4}$$

WORKED EXAMPLE

A specimen of insulating material has an electric susceptibility of $\chi_e = 6$, a cross-sectional area of $A = 6.45 \times 10^{-4}$ m², and a length of $d = 2 \times 10^{-3}$ m. If a voltage of $v = 10$V is applied across the material, calculate (1) the charge q induced on the surface, and (2) the polarization P.

Solution

The polarization P is given by

$$P = \varepsilon_o \chi_e E = \frac{\varepsilon_o \chi_e V}{d} \tag{5.5}$$

$$P = (8.85 \times 10^{-12}) \cdot (6) \cdot (10)/(2 \times 10^{-3}) \tag{5.6}$$

$$P = 2.67 \times 10^{-7} \; C \cdot m^{-2}. \tag{5.7}$$

TABLE 5.1
Dielectric Coefficients and Strengths for Some Dielectric Materials

Material	Dielectric Constant		Dielectric Strength (V/mil)[a]
	60 Hz	1 MHz	
Ceramics			
Titanate ceramics	—	15–10,000	50–300
Mica	—	5.4–8.7	1000–2000
Steatite (M_gO-SiO_2)	—	5.5–7.5	200–350
Soda-lime glass	6.9	6.9	250
Porcelain	6.0	6.0	40–400
Fused silica	4.0	3.8	250
Polymers			
Phenol-formaldehyde	5.3	4.8	300–400
Nylon 6,6	4.0	3.6	400
Polystyrene	2.6	2.6	500–700
Polystylene	2.3	2.3	450–500
Polyterafluoroethylene	2.1	2.1	400–500

[a] One mil = 0.001 in. These values of dielectric strength are average ones, the magnitude being dependent on specimen thickness and geometry, as well as the rate of application and duration of the applied electric field.

The charge q is given by

$$q = P \cdot A = (2.67 \times 10^{-7}) \cdot (6.45 \times 10^{-4}) \tag{5.8}$$

$$q = 1.71 \times 10^{-10} \, C. \tag{5.9}$$

5.1.4 VALUES OF DIELECTRIC COEFFICIENTS

The relative dielectric permittivities, $\varepsilon_r = \chi_e + 1$ of most materials are of order of magnitude 1–10, but in some materials, notably ferroelectrics, this value can be much higher, running into the range of 10,000. Some values of ε_r, the relative dielectric permittivity, are shown in Table 5.1

In fact, variation in the dielectric coefficient (permittivity) across the surface of a material can be used in materials evaluation as described by Han and Zhang [1].

5.2 ELECTRICAL CONDUCTORS

When an electrical conductor is subjected to an electric field, it results in the flow of an electric current.

5.2.1 ELECTRIC CURRENT AND CURRENT DENSITY

An electric current is the rate of passage of electric charge:

$$i = \frac{q}{t}. \tag{5.10}$$

A current of 1 A is equivalent to the passage of 1 C/sec.

The current density is the current passing per unit area through a conductor:

$$J = \frac{i}{A}. \tag{5.11}$$

5.2.2 RELATIONSHIP BETWEEN CURRENT DENSITY, CONDUCTIVITY, AND ELECTRIC FIELD

As shown in Figure 5.2, the current passing through a conductor is proportional to the electric field strength and the conductivity, so that

$$J = \frac{i}{A} = \sigma E. \tag{5.12}$$

The values of the electrical conductivity of different metals are shown in Table 5.2. The conductivity of a material is structure sensitive, so that the buildup of defects causes a reduction in conductivity.

5.2.3 MOVEMENT OF ELECTRONS IN CONDUCTING MATERIALS

When electrons move through a material, they do not move in straight lines over extended distances, rather, they are scattered as they move, as shown in Figure 5.3.

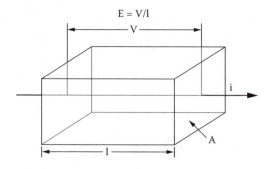

FIGURE 5.2 Effect of an electric field on the conductivity of a material.

TABLE 5.2
Room-Temperature Electrical
Conductivities for Some Common
Metals and Alloys

Metal	Electrical Conductivity $[(\Omega\text{–m})^{-1}]$
Silver	6.8×10^7
Copper	6.0×10^7
Gold	4.3×10^7
Aluminum	3.8×10^7
Iron	1.0×10^7
Brass (70 Cu-30 Zn)	1.6×10^7
Platinum	0.94×10^7
Plain carbon steel	0.6×10^7
Stainless steel	0.2×10^7

This means that more defects in the material result in more scattering and, therefore, there is a relation between defect density and electrical conductivity.

WORKED EXAMPLE

A metal wire 1 mm in diameter by 1 m in length is placed in an electric circuit. The voltage drop is 432 mV and it carries a current of 10 A. Calculate the electrical conductivity σ.

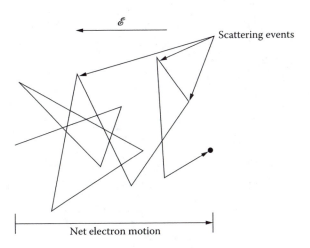

FIGURE 5.3 Movement of electrons through a material.

Solution

$$V = 432 \text{ mV}, L = 1 \text{ m} \tag{5.13}$$

$$E = V/L = 0.432 \text{ V} \cdot \text{m}^{-1} \tag{5.14}$$

$$A = 79 \times 10^{-8} \text{ m}^2, i = 10 \text{ A} \tag{5.15}$$

$$J = i/A = 12.73 \times 10^6 \text{ A} \cdot \text{m}^{-2} \tag{5.16}$$

$$\sigma = J/E = 12.73 \times 10^6 \text{ A} \cdot \text{m}^{-2}/432 \text{ mV} \tag{5.17}$$

$$\sigma = 29.5 \times 10^6 \text{ ohm}^{-1} \cdot \text{m}^{-1} \tag{5.18}$$

5.2.4 TEMPERATURE DEPENDENCE OF RESISTIVITY

The resistivity of most metals increases with temperature in an approximately linear fashion, as shown in Figure 5.4

The temperature dependence of the resistivity $\rho(T)$ can therefore be expressed as

$$\rho = \rho_o + \alpha T, \tag{5.19}$$

where ρ_o is the intrinsic resistivity term that is temperature independent and depends on scattering of electrons by the defects in the crystal structure of the material. The temperature-dependent term is the contribution to resistivity caused by scattering of electrons by phonons in the material.

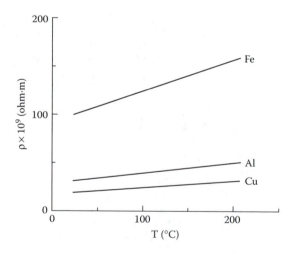

FIGURE 5.4 Variation of electrical resistivity with temperature.

The temperature-independent term can itself be split into contributions to resistivity due to the crystal lattice ρ_l, impurities and defects ρ_i, and stress and deformation ρ_σ.

$$\rho_o = \rho_l + \rho_i + \rho_\sigma. \tag{5.20}$$

This is Matthiessen's rule for addition of resistivities.

If the resistivity increases with temperature, the conductivity decreases and, in this case, the relationship is obtained directly from the above equation so that,

$$\frac{1}{\sigma} = \frac{1}{\sigma_o} + \alpha T, \tag{5.21}$$

or equivalently,

$$\sigma = \frac{\sigma_o}{1 + \alpha \sigma_o T}, \tag{5.22}$$

where σ_o is the intrinsic or temperature-independent term.

Resistivity also changes with alloy composition, because both ρ_l and ρ_i will change with composition. The change in resistivity with composition of various copper alloys is shown in Figure 5.5

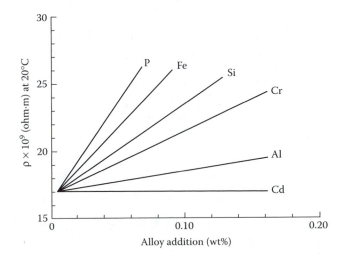

FIGURE 5.5 Dependence of ρ on composition for CuP, CuFe, CuSi, CuCr, CuAl, and CuCd alloys.

5.3 ELECTRICAL MEASUREMENTS FOR MATERIALS TESTING

Conductivity is dependent on structure and composition of the materials, which is an important result. It means that information about structure and/or composition is present in the resistivity data. Also, any other measurement that depends on conductivity must contain such information and, in principle, can be used to detect the state of a material. The main electrical nondestructive evaluation (NDE) method is not DC conductivity measurement, instead, it is the more practical one of eddy current inspection [2,3,4]. In this case, the response of the eddy-current sensor depends, in part, on the conductivity and permeability of the test material, and therefore on the presence of defects or discontinuities in the material [5].

As an example, when a current is passed through a test material using two electrodes, an electrical potential is created between the electrodes, and this electrical potential varies from location to location in an inhomogeneous material. Variation in the electric potential can therefore be used for materials evaluations as described by Gille [6].

5.3.1 GENERATION OF EDDY CURRENTS

The quantitative description of eddy-current effects relies heavily on mathematical formalisms. To follow these, we need to know some simple conventions in vector calculus. The vector differential operator, known as *del*, is a generalized differential having the form

$$\nabla = \frac{\partial}{\partial x}\underline{i} + \frac{\partial}{\partial y}\underline{j} + \frac{\partial}{\partial z}\underline{k}. \tag{5.23}$$

There can be a defined scalar product and a vector product, in accordance with the normal expressions used in vector algebra; so that if $x = a\underline{i} + b\underline{j} + c\underline{k}$, the scalar product is

$$\nabla \cdot \underline{x} = \frac{\partial a}{\partial x} + \frac{\partial b}{\partial y} + \frac{\partial c}{\partial z}, \tag{5.24}$$

and the vector product is

$$\nabla \times \underline{x} = \left(\frac{\partial c}{\partial y} - \frac{\partial b}{\partial z}\right)\underline{i} + \left(\frac{\partial a}{\partial z} - \frac{\partial c}{\partial x}\right)\underline{j} + \left(\frac{\partial b}{\partial x} - \frac{\partial a}{\partial y}\right)\underline{k}. \tag{5.25}$$

A time-varying magnetic flux density B will cause a circulating electric field, according to Faraday's law,

$$\nabla \times E = -\frac{\partial B}{\partial t}, \tag{5.26}$$

which causes an electrical current in a conducting material when it is exposed to the induced electric field.

5.3.2 Penetration of a Plane Electromagnetic Wave into a Material

The equation for the penetration depth is obtained by solving the wave equation for a plane electromagnetic wave impinging on a flat surface. The equation for a time-varying magnetic field is [7]

$$\nabla^2 H - \varepsilon \mu_o \mu_r \frac{\partial^2 H}{\partial t^2} - \sigma \mu_o \mu_r \frac{\partial H}{\partial t} = 0, \tag{5.27}$$

and supposing that the magnetic field is sinusoidal,

$$H = H_o e^{i\omega t}. \tag{5.28}$$

Substituting this into the field equation gives

$$\nabla^2 H - \omega^2 \varepsilon \mu_o \mu_r H - i\omega \sigma \mu_o \mu_r H = 0. \tag{5.29}$$

Supposing there is a plane wave

$$\nabla^2 H = \frac{\partial^2 H}{\partial x^2}, \tag{5.30}$$

this then gives,

$$\frac{\partial^2 H}{\partial x^2} - \omega^2 \varepsilon \mu_o \mu_r H - i\omega \sigma \mu_o \mu_r H = 0. \tag{5.31}$$

There is a solution of this equation for H which is

$$H = H_o e^{\left(\left(\alpha + \frac{i}{\delta} \right) x \right)}, \tag{5.32}$$

which solves the previous equation when

$$\delta = \left(\frac{1}{\pi \nu \sigma \mu} \right)^{1/2}. \tag{5.33}$$

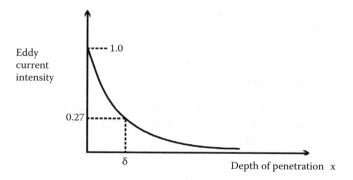

FIGURE 5.6 Decay of intensity of eddy currents with depth inside a material.

This gives the depth at which the value of H decays to $1/e$ of its value at $x = 0$.

Therefore, the often quoted penetration depth formula is only really applicable in the special case of a plane wave impinging on a flat surface. The variation of the intensity of eddy currents with depth is shown in Figure 5.6

5.3.3 SKIN DEPTH

The rate of decay of the eddy-current intensity with depth is also usually represented by the penetration depth (or skin depth). This represents the depth at which the intensity of the eddy currents has fallen to $1/e$ times its value at the surface of the material [8]. In the specific case of a plane electromagnetic wave impinging on a flat surface, the dependence of the penetration depth on frequency, conductivity and permeability follows the relation

$$\delta = \sqrt{\frac{2}{\omega \mu_r \mu_o \sigma}}. \tag{5.34}$$

WORKED EXAMPLE

Calculate the penetration depth, δ, for steel with conductivity of 1.5×10^6 S $\cdot$ m^{-1} and a permeability of 1.3×10^{-6} H $\cdot$ m^{-1} at a frequency of 0.16 MHz.

Solution

Assuming a plane wave is impinging on a flat surface, the classical skin-depth equation can be used. Using this equation to give an approximation to the skin depth, and replacing ω with the linear frequency ν,

$$\delta = \sqrt{\frac{1}{\pi \nu \mu_r \mu_o \sigma}}, \tag{5.35}$$

and substituting in the values gives

$$\delta = \sqrt{\frac{1}{0.98 \times 10^6}} = \sqrt{1.02 \times 10^{-6}} \qquad (5.36)$$

$$\delta = 1.01 \times 10^{-3} \text{ m.} \qquad (5.37)$$

5.3.4 ELECTRICAL PARAMETERS

Here are the electrical parameters of interest in characterizing the electrical response of the sensor coil in eddy-current inspection

$$\textit{Resistance} \quad R = \frac{V}{i}, \qquad (5.38)$$

and for a resistive element made of a material of resistivity ρ, with length l and cross-sectional area A, the resistance is given by

$$R = \frac{\rho l}{A}. \qquad (5.39)$$

$$\textit{Inductance} \quad L = \frac{V}{\left(\dfrac{di}{dt}\right)}, \qquad (5.40)$$

and for a coil of N turns with length l and cross-sectional area A, which is filled with a material of relative permeability μ_r

$$L = \frac{\mu_o \mu_r N^2 A}{l}. \qquad (5.41)$$

$$\textit{Capacitance} \quad C = \frac{Q}{V} \qquad (5.42)$$

$$\textit{Reactance} \quad X = 2\pi v L - \frac{1}{2\pi v C} \qquad (5.43)$$

$$\textit{Impedance} \quad Z = \sqrt{(R^2 + X^2)} \qquad (5.44)$$

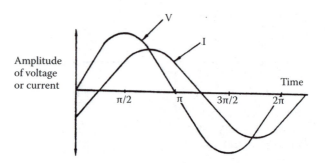

FIGURE 5.7 Time dependence of a sinusoidal voltage and the corresponding sinusoidal current, showing that the peak in current occurs after the peak in voltage. This is known as a "phase lag."

5.3.5 RELATIONSHIP BETWEEN VOLTAGE AND CURRENT UNDER AC EXCITATION

As a result of the impedance having two components, the voltage and current of a circuit containing both resistive and reactive components are not usually in phase. This is shown in Figure 5.7. The time lag of the current behind the voltage is known as the phase angle ϕ.

The response characteristics of the test material depend on the mutual inductance between the coil and the test specimen. The general relationship between current and voltage is in the form of a differential equation, which can be applied to DC voltages, sinusoidal AC voltages, and even nonsinusoidal AC voltages.

$$V = L\frac{di}{dt} + iR + C\int idt. \tag{5.45}$$

Ignoring capacitance, which is usually a negligible quantity in these cases, the voltage–current relationship can also be expressed equivalently by the differential equation

$$V = L\frac{di}{dt} + iR. \tag{5.46}$$

The general relationship between voltage and current is through the impedance Z, which is obtained as a solution to the above differential equation

$$V = iZ, \tag{5.47}$$

and, in this way, the impedance of the eddy-current sensors can be used to evaluate the condition of the test material through a process of calibration.

5.3.6 COMPONENTS OF IMPEDANCE

For a sinusoidal ac current

$$i = i_0 (\cos(\omega t - \phi) + j \sin(\omega t - \phi)) \qquad (5.48)$$

where ϕ is the phase angle relative to the applied voltage and $j = \sqrt{-1}$. Therefore

$$L \frac{di}{dt} = \omega L i_0 (-\sin(\omega t - \phi) + j \cos(\omega t - \phi)) = j\omega L i_0 (\cos(\omega t - \phi) + j \sin(\omega t - \phi))$$

$$(5.49)$$

$$iR = i_0 R(\cos(\omega t - \phi) + j \sin(\omega t - \phi)) \qquad (5.50)$$

Substituting this into the equation (5.46) for voltage gives

$$V = i(R + j\omega L) \qquad (5.51)$$

so replacing $j\omega L$ with the reactance X_L, the impedance Z is

$$Z = R + X_L \qquad (5.52)$$

which can be displayed as a vector on a two-dimensional impedance plane for ease of presentation and interpretation. In these cases the impedance vector has two orthogonal components, the real component (resistance), and the imaginary component (reactance). It is easy to show from the above equations that $\tan \phi = \frac{\omega L}{R}$.

5.4 MAGNETIC FIELDS

A magnetic field is created whenever there is charge in motion. This means that when current passes through a conductor, a magnetic field is always produced. However, magnetic fields are also produced in situations where there is no conventional electric current, but in which there is still charge in motion. The most familiar example is the production of a magnetic field by a permanent magnet. The response in terms of the magnetic flux density in the material is dependent on the strength of the magnetic field and the permeability of the material [7]. Magnetic fields can be measured in a variety of ways, most commonly by use of a gaussmeter or a fluxmeter [9].

Magnetic fields affect materials. The most dramatic examples occur when a ferromagnet, such as iron, nickel, or cobalt, or one of their alloys, is exposed to a magnetic field. Time-dependent magnetic fields can cause currents to flow in electrically conducting materials, as we have seen above in the section discussing eddy currents. The strength of the eddy currents depends on the conductivity and the permeability of the material. Therefore, we have two structure-sensitive materials properties that can be probed by magnetic field techniques — permeability and conductivity.

5.4.1 MAGNETIC FIELD H

A magnetic field is produced whenever there is an electric current. As a simple example, when current i passes down a long straight wire, there is a magnetic field which circulates around the wire. At any radial distance r, the field strength is

$$H = \frac{i}{2\pi r}. \tag{5.53}$$

5.4.2 MAGNETIC INDUCTION B

The magnetic induction B is the result of the presence of a magnetic field H. This can be interpreted as the response of a material to the applied magnetic field. The magnetic induction B is related to magnetic field H through the permeability, μ, so that in free space

$$B = \mu_o H, \tag{5.54}$$

whereas in materials

$$B = \mu_o \mu_r H. \tag{5.55}$$

μ_o is the permeability of free space, which has a fixed value of $4\pi \times 10^{-7}$ H/m, and μ_r is the relative permeability of the material, which can vary from $\mu_r = 1$ in nonmagnetic materials, to 10^7 in ultrahigh permeability magnetic metallic glasses. From this, it can be seen that the relative permeability of free space is also 1, so that it becomes apparent that nonmagnetic materials have little effect on the magnetic flux density (Figure 5.8).

   If the relative permeability μ_r is known, then the magnetic induction B can be calculated for a given field strength.

WORKED EXAMPLE

A magnetic material is exposed to a field of 1000 A/m and this causes a magnetization of 0.5×10^6 A/m. Calculate (1) the magnetic induction B, (2) the permeability, (3) the susceptibility χ, and (4) the relative permeability.

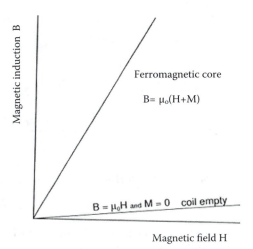

FIGURE 5.8 Variation of magnetic induction with magnetic field strength in the case of an idealized "linear" material.

Solution

1. The equation linking magnetic induction with magnetization and magnetic field strength is

$$B = \mu_o(H + M) = (12.56 \times 10^{-7})(500 \times 10^3 + 1 \times 10^3) = 0.629 \text{ T}. \qquad (5.56)$$

2. $$\mu = \frac{B}{H} = \frac{0.629}{1000} = 0.629 \times 10^{-3} \text{ H/m} \qquad (5.57)$$

3. $$\chi = \frac{M}{H} = \frac{500 \times 10^3}{1000} = 500 \qquad (5.58)$$

4. $$\mu_r = \frac{\mu}{\mu_o} = \frac{0.629 \times 10^{-3}}{12.56 \times 10^{-7}} = 501 \qquad (5.59)$$

The only materials property that appears in the constitutive equation relating B and H is μ_r. This is a structure-sensitive property of the material, the measurement of which can give information on the state of the material. In strongly magnetic materials, such as a ferromagnet like iron, nickel, cobalt, or their alloys steel, the permeability is not constant, and so B is a nonlinear function of H [9]. The nonlinearity can be represented by a saturable curve of the form shown in Figure 5.9. This can be described by a saturable function. The exact detail of this nonlinear relationship is dependent on structure, so that the measurement of the dependence of B on H can be used for nondestructive evaluation [10].

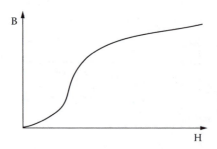

FIGURE 5.9 Nonlinear variation of B with H is structure sensitive and can be used for nondestructive evaluation.

More exactly, B is a hysteretic function of H. This is much harder to describe mathematically than the nonlinear relationship in the previous figure.

5.4.3 MAGNETIC FIELDS IN VARIOUS CONFIGURATIONS

A linear current passing down a linear conductor results in a magnetic field, which circulates around the conductor, as shown in Figure 5.10

We could say that the magnetic field "curls" around the current. This can be expressed mathematically as the curl of the magnetic field, which equals the current density

$$\nabla \times H = \frac{i}{A}. \tag{5.60}$$

For a circulating current, such as the current passing around a single loop of conductor as shown in Figure 5.11, the magnetic field is linear.

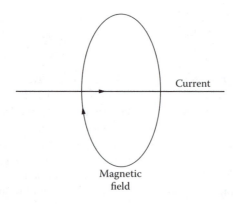

FIGURE 5.10 Magnetic field of a linear conductor carrying a current.

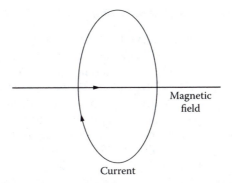

FIGURE 5.11 Magnetic field of a single circular loop of conductor carrying an electric current.

5.4.4 THREE SIMPLE CASES

There are three particularly simple cases of the relationship between field and current. The single circular loop of radius a, carrying a current i, produces a field at the center of the coil given by

$$H = \frac{i}{2a}. \tag{5.61}$$

In the case of a linear conductor carrying a current of i amperes, the field strength circulating at a radial distance a is,

$$H = \frac{i}{2\pi a}. \tag{5.62}$$

In the case of a "long" solenoid containing n turns per unit length, the field at the center of the solenoid is

$$H = ni. \tag{5.63}$$

For a solenoid with number of turns N and length ℓ, the field at the center is given approximately by

$$H \approx \frac{Ni}{\ell} \tag{5.64}$$

where the approximation improves as ℓ becomes larger.

5.4.5 LEAKAGE FLUX IN THE VICINITY OF FLAWS

Large values of B imply large leakage fields in the vicinity of cracks or flaws in a material. Therefore, by magnetizing a part and measuring the induced field B, the surface flaws can be detected, located, and approximately sized. Both DC fields [8] and AC fields [11] have been used for flaw detection in ferrous components. Exact sizing of flaws by this method can still be problem because, although the effect of a particular flaw on the magnetic field in its vicinity can be calculated, the inverse procedure — the calculation of the flaw profile from the leakage field — does not necessarily have a unique solution.

The leakage flux is determined not only by the size of the flaw, but also by the magnetic properties of the test material [12]. These magnetic properties are most often represented by a BH curve, as shown in Figure 5.9, although in practice, the magnetization curves are more complicated than this because of hysteresis. An alternative method for detecting leakage flux and, hence, inhomogeneities in materials due to defects, is through the use of magnetooptic imaging, whereby an optical sensor is used in which the direction of polarization of light is altered by the presence of the magnetic fields at the surface of the test material [13].

EXERCISES: ELECTRICAL AND MAGNETIC PROPERTIES

5.1 Explain how eddy currents are generated in a material. Calculate the frequency of excitation required to achieve a depth of penetration of 3 mm in a test specimen with permeability of 3.5×10^{-4} H/m and conductivity of 5×10^5 S/m.

5.2 An eddy current probe has a DC resistance in air of $R = 10$ ohms and an inductance of $L = 125 \times 10^{-6}$ H. To operate correctly at 400 Hz, it needs to have a total impedance of $Z = 12$ ohms. Find the value of the tuning capacitor C that you need to add to give the probe the necessary total impedance. Then, calculate the standard depth of penetration δ in a piece of aluminum having a permeability of $4\pi \times 10^{-7}$ H/m and conductivity of 35×10^6 S/m.

5.3 A long solenoid with 10 turns per millimeter is to be used to magnetize two steel rods for magnetic particle inspection. One has a relative permeability of 20, and the other has a relative permeability of 75. If a flux density of 1.1 T is optimal for magnetic particle inspection, calculate the current that needs to be passed through the coils of the solenoid in each case. If the required magnetization for magnetic particle inspection is to be achieved by passing an axial electric current along the center of the rods, and both have a diameter of 10 mm, calculate the current that is needed to produce the appropriate flux density on the surface of the parts.

REFERENCES

1. Han, H.C. and Zhang, J., Extraction of permittivity from reflection data: lossless case, *Research in NDE* 11, 25, 1999.
2. Libby, H.L., *Introduction to Electromagnetic Nondestructive Evaluation*, John Wiley & Sons, New York, 1971.
3. Libby, H.L., Basic principles and techniques of eddy current testing, *NDT* 14, 12, 1956.
4. Blitz, J., *Electrical and Magnetic Methods of Nondestructive Testing*, 2nd ed., Chapman and Hall, London, 1997, 261 pp.
5. R.C. McMaster, *The Present and Future of Eddy Current Testing*, *Mater Eval* 60, 27, January 2002.
6. Gille, G., The electrical potential method and its application to non-destructive testing, *Nondestr Test Eval Int* 4, 36, 1971.
7. Jiles, D.C., *Introduction to Magnetism and Magnetic Materials*, 2nd ed., Chapman and Hall, London, 1998.
8. Franklin, E.M., Eddy current inspection, *Mater Eval* 40, 1008, 1982.
9. Stanley, R.K., Magnetic field measurement: the gauss meter in magnetic particle testing, *Mater Eval* 46, 1509, November 1988.
10. Paul I. Nippes and Elizabeth N. Galano, How and why to measure magnetism accurately, *Mater Eval* 60, 507, April 2002.
11. Raine, A. and Cameron, R., Alternating current field measurement, *Mater Eval* 60, 389, 2002.
12. Katoh, M., Nishio, K., and Yamaguchi, T., The influence of modeled B-H curve on the density of the magnetic leakage flux due to a flaw using yoke-magnetization, *Nondestr Test Eval Int* 37, 603, 2004.
13. Novotny, P., Sajdl, P., and Machac, P., A magneto-optic imager for NDT applications, *Nondestr Test Eval Int* 37, 645, 2004.

Further Reading

Lord, W. (Ed.), *Electromagnetic Methods of Nondestructive Testing*, Gordon and Breach, New York, 1985.

http://www.ndt-ed.org/EducationResources/CommunityCollege/EddyCurrents/cc_ec_index.htm.

http://www.ndt-ed.org/EducationResources/CommunityCollege/MagParticle/cc_mpi_index.htm.

6 Effects of Radiation on Materials

This chapter looks at the interactions of ionizing radiation with materials, including the basic equations used to quantify the variation of radiation intensity in materials, and how these equations can be developed from an understanding of the basic interactions of radiation with atoms in the material. In addition to the use of radiation for bulk interrogation of materials, a summary of the use of radiation for surface characterization is given, particularly for determination of surface chemistry. Finally, the chapter includes a discussion of concepts of exposure dose and dose rate, and how these are measured and expressed in quantitative terms.

6.1 BASICS OF X-RAYS

6.1.1 GENERATION OF X-RAYS

X-rays are produced when high-energy electrons are emitted from a source known as a filament and collide with a metal target, usually made from tungsten, copper, or gold. The electrons excite the atoms in the target, elevating other electrons to higher energy levels. When these electrons revert to their original states, an x-ray is emitted [1]. As a result, a spectrum of x-ray energies is emitted, rather than a "monochromatic" x-ray beam. The efficiency of this process is about 1% at the lower-energy ranges, but can be as high as 25% at the higher energies.

Figure 6.1 shows an x-ray source consisting of the filament at the top, a set of equipotential rings to accelerate the electrons once they are liberated from the filament, and a target anode at the bottom.

The source of x-ray has a finite size which, in effect, is the projected area of the target along the direction of the diverging x-ray beam. Portable x-ray sets are important for the use of x-rays outside the laboratory, and these devices have been readily available since the 1970s [2]. X-rays are detected by using an x-ray sensitive film, gas-filled detectors (such as Geiger tubes, proportional counters, and ion chambers), or by thermoluminescent detectors [3].

6.1.2 TYPICAL X-RAY SPECTRUM

The energy of the x-ray beam is often quoted as a single number, usually in electron volts. However, as mentioned above, there is, in practice, a spectrum of energies. The number that is usually quoted for convenience is the energy of the peak intensity of the x-rays. This is shown in Figure 6.2. The energy of the x-rays is not the quoted

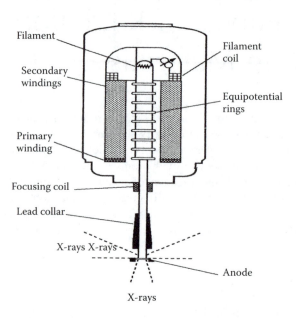

FIGURE 6.1 An X-ray source showing the filament, equipotential rings for accelerating the electrons, and a coil to provide magnetic focusing of the electron beam.

energy of the x-ray set in terms of a direct conversion of the voltage of the set into eV. In fact, the peak energy of the x-ray spectrum is usually about two thirds of the voltage of the x-ray set, so that a 300-kV set gives an x-ray spectrum with a peak at about 200 keV.

6.1.3 ATTENUATION OF RADIATION

Several independent mechanisms are responsible for the decrease in intensity of radiation (attenuation) as it passes through materials. On the macroscopic or continuum scale, these processes can be classified as either absorption processes or scattering processes. In most cases, both occur simultaneously.

For a nondiverging beam of radiation, the decrease in intensity as it passes through a material is expected to be exponential, provided no other causes of attenuation of the signal are at work. Therefore, the following equation can often be applied [4, p. 14],

$$I(x) = I(0) \cdot \exp(-\mu x), \tag{6.1}$$

where $I(x)$ is the intensity of radiation at a distance x, $I(0)$ is the incident intensity, and μ is the attenuation coefficient, which has units of reciprocal distance.

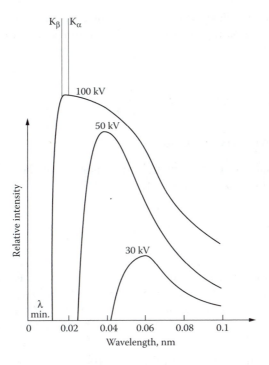

FIGURE 6.2 Typical X-ray spectrum showing the range of energies and, within this, a peak energy, which is usually quoted as the energy of the X-rays.

This equation is only true when the beam of radiation is not diverging; sometimes, this is referred to as a collimated (meaning nondivergent) source.

6.1.4 ATTENUATION COEFFICIENTS

There are a few ways of classifying the attenuation. The most obvious is the linear attenuation coefficient μ, as given in the above equation, which can be calculated from measurements of the decrease in intensity of radiation over a known distance:

$$\mu = \frac{1}{x} \log_e \left(\frac{I(0)}{I(x)} \right). \tag{6.2}$$

The half-value thickness (HVT) is a commonly used practical measure of the attenuation of radiation. This is the depth of material that is needed to cause the intensity of radiation to decrease to half of its value on the surface:

$$HVT = \frac{\log_e 2}{\mu}. \tag{6.3}$$

The tenth-value thickness (TVT) is a similar concept, except that it is the depth needed to cause the intensity of radiation to decrease to one-tenth of its value on the surface,

$$TVT = \frac{\log_e 10}{\mu}.$$ (6.4)

6.1.5 MASS ATTENUATION COEFFICIENT

The mass attenuation coefficient μ_m is the quotient of the linear attenuation coefficient and the density of the material [5, p. 72],

$$\mu_m = \frac{\mu}{\rho},$$ (6.5)

and the units of μ_m are necessarily $cm^2.g^{-1}$ or $m^2.kg^{-1}$. Table 6.1 shows the mass absorption/attenuation coefficients for all elements for x-ray wavelengths of 0.056, 0.071, and 0.154 nm. By combining this information with the density of materials, the linear absorption coefficients can be calculated.

6.1.6 COMPOSITE ATTENUATION COEFFICIENTS

If there are several components present in a material with mass fractions of $f_1 = m_1/m_{tot}, f_2 = m_2/m_{tot}, f_3 = m_3/m_{tot}, \ldots f_n = m_n/m_{tot}$, each of which has a different mass attenuation coefficient, $\mu_1, \mu_2, \mu_3, \ldots, \mu_n$, then the overall or average mass attenuation coefficient can be determined from the individual mass attenuation coefficients by the equation,

$$\mu_m = f_1\mu_1 + f_2\mu_2 + f_3\mu_3 + \quad + f_n\mu_n.$$ (6.6)

This simply means that the overall mass attenuation coefficient of the material is the weighted sum of the mass attenuation coefficient of the various components.

WORKED EXAMPLE

Calculate the x-ray attenuation for $BaCO_3$, given that the mass attenuation coefficient of the components is:

Element	Atomic Weights	Mass Attenuation Coefficient
Ba	137.4	358.9
C	12	5.50
O	16	12.7

TABLE 6.1
Mass absorption coefficients for x-rays of wavelength $\lambda = 0.56$, 0.71 and 1.54 Å

	Mass Absorption Coefficient (μ_m), cm²/g				Mass Absorption Coefficient (μ_m), cm²/g		
Absorber	Ag K_α 0.56 Å	Mo K_α $\lambda = 0.71$ Å	Cu K_α $\lambda = 1.54$ Å	Absorber	Ag K_α 0.56 Å	Mo K_α $\lambda = 0.71$ Å	Cu K_α $\lambda = 1.54$ Å
H	0.371	0.3727	0.435	Zr	58.5	16.10	143
Li	0.187	0.1968	0.716	Nb	61.7	16.96	153
Be	0.229	0.2451	1.50	Mo	64.8	18.44	162
B	0.279	0.3451	2.39	Pd	12.3	24.42	206
C	0.400	0.5348	5.50	Ag	13.1	26.38	218
N	0.544	0.7898	7.52	Cd	14.0	27.73	231
O	0.740	1.147	12.7	In	14.9	29.13	243
F	0.976	1.584	16.4	Sn	15.9	31.18	256
Na	1.67	2.939	30.1	Sb	16.9	33.01	270
Mg	2.12	3.979	38.6	Te	17.9	33.92	282
Al	2.65	5.043	48.6	I	19.0	36.33	294
Si	3.28	6.533	60.6	Cs	21.3	40.44	318
P	4.01	7.870	74.1	Ba	22.5	42.37	358.9
S	4.84	9.625	89.1	La	23.7	45.34	341
Cl	5.77	11.64	106	Ce	25.0	48.56	352
K	8.00	16.20	143	Pr	26.3	50.78	363
Ca	9.28	19.00	162	Nd	27.7	53.28	374
Sc	10.7	21.04	184	Sm	30.6	57.96	397
Ti	12.3	23.25	208	Gd	33.8	62.79	437
V	14.0	25.24	233	Tb	35.5	66.77	273
Cl	15.8	29.25	260	Dy	37.2	68.89	286
Mn	17.7	31.86	285	Er	40.8	75.61	134
Fe	19.7	37.74	308	Yb	44.8	80.23	146
Co	21.8	41.02	313	Hf	48.8	86.33	159
Ni	24.1	47.24	45.7	Ta	50.9	89.51	166
Cu	26.4	49.34	52.9	W	53.0	95.76	172
Zn	28.8	55.46	60.3	Re	55.2	98.74	178
Ga	31.4	56.90	67.9	Os	57.3	100.2	186
Ge	34.1	60.47	75.6	Ir	59.4	103.4	193
As	36.9	65.97	83.4	Pt	61.4	108.6	200
Se	39.8	68.82	91.4	Au	63.1	111.3	208
Rb	48.9	83	117	Hg	64.7	114.7	216
Sr	52.1	88.04	125	Pb	67.7	122.8	232
Y	55.3	97.56	134	Bi	69.1	125.9	240

Solution

Using the above equation for the mass absorption coefficient of a material consisting of several distinct individual chemical components,

$$\mu_m = f_1\mu_1 + f_2\mu_2 + f_3\mu_3 \tag{6.7}$$

$$\mu_m = \frac{137.4}{197.4}358.9 + \frac{12}{197.4}5.5 + \frac{48}{197.4}12.7 = 253 \text{ cm}^2/\text{g}. \tag{6.8}$$

The attenuation coefficient for $BaCO_3$ is therefore 254 cm²/g. This number is determined from the absorption coefficients of its chemical components, so that it does not depend at all on the structure of the material or any processing parameters.

Of course, we are not limited to the use of x-rays, although on the whole, they are often the most convenient form of penetrating radiation for use in nondestructive testing. A closely related alternative is the use of gamma radiation. The gamma rays are generated by radioactive materials in which an unstable nucleus de-excites with emission of electromagnetic radiation, in this case, a gamma ray. Some additional safety requirements are needed when using gamma ray sources, because gamma rays originate in unstable radioactive nuclei. Therefore, unlike x-rays which are generated by an x-ray set that can be switched off, the source of gamma rays cannot be so easily controlled [6]. The detection methods for gamma rays are identical to those used for x-rays.

6.2 INTERACTION OF X-RAYS WITH MATERIALS

We first ask, how does radiation interact with materials? Later we ask, how does radiation alter materials? The second question is important if we want to use radiation for nondestructive evaluation of materials, because such interactions must not alter the material.

There are a few different mechanisms whereby these interactions take place [5, pp. 64–72]. The probability of interaction of an x-ray with an atom can be expressed in terms of a cross-section, and this atomic scale cross-section is related to the macroscopic attenuation or absorption coefficients described above. The penetration depth of x-rays into matter is also related to these atomic scale properties.

6.2.1 PRINCIPAL INTERACTION PROCESSES

The main physical mechanisms of interaction are as follows:

1. Classical Thompson scattering: This is an elastic scattering process, which means that the x-ray photon interacts with a whole atom, is scattered, and there is no change in energy of the photon. A simplistic

way of thinking about this is to suppose that the photon is merely diverted, although this interpretation of the event is not strictly correct.

2. Incoherent Compton scattering: This is an inelastic process whereby the x-ray interacts with an electron, is absorbed, and another x-ray is emitted with lower energy. The electron then recoils after collision with the photon. Momentum is conserved, and the recoil energy of the electron is equal to the difference in energies between the incident and emitted x-ray photons.

3. Photoelectric effect: In the photoelectric effect, the x-ray is absorbed by the material and an electron is ejected. This requires a threshold energy for the event to occur, and all photon energy is absorbed in the process. This process is the inverse of that used to generate the x-ray in the target.

4. Pair production: In this process, the photon is destroyed and its energy is converted into an electron–positron pair. This ensures that charge neutrality is maintained. By conservation of energy, a threshold energy for this process is required that is equal to the rest energy of the two particles that are produced, meaning that pair production does not occur for photons with energy less than 1.02 MeV.

5. Photo disintegration: In the process of photo disintegration, an x-ray strikes the nucleus and liberates a subnuclear particle from the nucleus. The incident photon must have sufficient energy to burrow its way right to the nucleus, where it strikes the nucleus and is destroyed. This requires very high photon energy, and so is a process that occurs with a low probability for typical x-ray energies that are used in NDE. It can therefore be considered to be of negligible practical significance for materials evaluation purposes.

All of the preceding processes contribute to an overall attenuation coefficient μ for radiation passing through a material. This attenuation coefficient varies for different types and energies of radiation and, of course, also varies from material to material [4, pp. 14–18].

6.2.2 ATOMIC ATTENUATION COEFFICIENT

The atomic attenuation coefficient is the same as the x-ray cross-section of the atom and represents the statistical decrease in intensity when an x-ray traverses an atom,

$$\mu_a = \frac{\mu}{\rho} \cdot \frac{A_w}{N_o}, \tag{6.9}$$

where A_w is the atomic mass and N_o is Avogadro's number. The units are m^2 per atom or cm^2 per atom

6.2.3 ELECTRONIC ATTENUATION COEFFICIENT

The scattering of x-ray is principally a process whereby the x-rays interact with electrons. Therefore, an electronic attenuation coefficient can be defined by

$$\mu_e = \frac{\mu}{\rho N_e}, \tag{6.10}$$

where N_e is the number of electrons contained in unit mass of the absorbing material. The units for this quantity are therefore m^2 per electron or cm^2 per electron.

6.2.4 CONTRIBUTIONS TO ATTENUATION

Each of the above processes contributes to the attenuation of radiation as it passes through a material, and so μ can be written in a compound form

$$\mu = \mu_t + \mu_c + \mu_{ph} + \mu_{pp}, \tag{6.11}$$

where the various terms in the equation are the attenuations due to Thompson scattering, Compton scattering, photoelectric effect, and pair production, respectively.

The relative contributions of the various mechanisms to the attenuation of radiation in materials changes depending on the energy of the radiation. This is shown in Figure 6.3. It can be seen that at low energies, the attenuation is dominated by the photoelectric effect and elastic Thompson scattering. Inelastic Compton scattering increases with x-ray energy at low energies, and reaches a peak at about 0.1 MeV. Pair production only begins to play a role above the threshold of 1.02 MeV. Above 1 MeV, the photoelectric effect is negligible, and above 5 MeV, coherent Thompson scattering is negligible.

6.2.5 ENERGY DEPENDENCE OF ATTENUATION COEFFICIENTS

As shown above, the mass attenuation coefficient, μ_m, differs from element to element and from material to material. The attenuation coefficient for a particular element or a particular material also changes with energy of the incident radiation. To deal with this situation, tables of values of the attenuation coefficients for different elements and materials at different energies are available, and the values of μ_m are obtained from these. As an example, mass absorption coefficients for x-ray energies from 20 keV to 20 MeV are shown in Table 6.2. On the whole, the absorption coefficient decreases with the energy of radiation, so that in a simplified sense, the more energetic particles are just more difficult to stop.

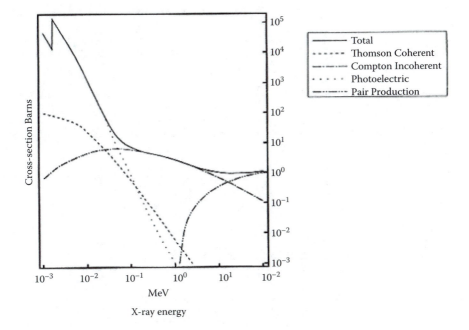

FIGURE 6.3 Energy dependence of the cross-section for x-ray attenuation of the principal interaction mechanisms of x-rays with aluminum.

WORKED EXAMPLE

An x-ray beam is needed that will transmit 5% intensity through 2.5 cm of steel. Determine the energy of the x-rays that are needed. The density of the steel is 7.87 gm/cm³, and the mass attenuation coefficients can be found from the previous table.

Solution

Because the intensity needs to be 5%,

$$\frac{I}{I_o} = \exp\left(-2.54\mu\right) = 0.05 \tag{6.12}$$

$$\mu = -\frac{\log_e(0.05))}{2.54} = 1.179. \tag{6.13}$$

$$\text{Therefore, } \mu_m = \frac{1.179}{7.87} = 0.15 \text{ cm}^2/\text{g}. \tag{6.14}$$

From the table of attenuation coefficients for iron, $\mu_m = 0.147$ cm²/g at an x-ray energy of 200 keV. Therefore, the required energy is 200 keV.

TABLE 6.2
Mass Absorption Coefficient cm²/g

Energy	Titanium	Vanadium	Chromium	Iron	Nickel	Copper	Zinc	Zirconium	Molybdenum	Silver
20 keV	15.72	17.06	19.83	25.72	32.33	33.86	38.11	71.58	81.05	18.05
40	2.174	2.369	2.767	3.555	4.487	4.779	5.347	11.46	13.09	17.19
60	0.758	0.821	0.947	1.176	1.460	1.562	1.726	3.721	4.331	5.764
80	0.405	0.432	0.489	0.548	0.707	0.752	0.820	1.692	1.993	2.648
100	0.274	0.288	0.319	0.367	0.433	0.455	0.490	0.944	1.116	1.469
200	0.132	0.133	0.139	0.147	0.158	0.157	0.162	0.220	0.249	0.300
400	0.0910	0.0899	0.0924	0.0941	0.0977	0.0944	0.0956	0.1016	0.1059	0.1136
600	0.0753	0.0742	0.0760	0.0770	0.0794	0.0763	0.0770	0.0774	0.0787	0.0814
800	0.0657	0.0646	0.0662	0.0669	0.0688	0.0660	0.0665	0.0658	0.0613	0.0674
1 MeV	0.0589	0.0579	0.0593	0.0599	0.0615	0.0589	0.0594	0.0580	0.0582	0.0589
2	0.0418	0.0411	0.0421	0.0426	0.0438	0.0420	0.0424	0.0414	0.0416	0.0420
4	0.0317	0.0314	0.0324	0.0331	0.0344	0.0322	0.0335	0.0345	0.0350	0.0360
6	0.0287	0.0274	0.0295	0.0304	0.0320	0.0311	0.0316	0.0338	0.0344	0.0361
8	0.0274	0.0271	0.0285	0.0298	0.0315	0.0305	0.0312	0.0343	0.0351	0.0372
10	0.0269	0.0276	0.0281	0.0296	0.0315	0.0306	0.0313	0.0352	0.0360	0.0385
20	0.0282	0.0304	0.0301	0.0319	0.0346	0.0339	0.0348	0.0410	0.0424	0.0459

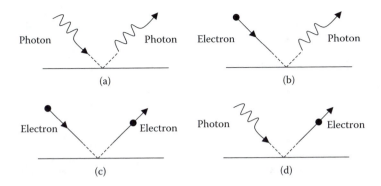

FIGURE 6.4 Summary of the related surface analysis techniques (a) XRF, (b) EDS, (c) AES, and (d) XPS, showing combinations of electrons/photons impinging on a surface and photons/ electrons being emitted.

6.2.6 SURFACE ANALYSIS USING RADIATIVE METHODS

In cases where chemical analysis of a material is required, sampling of the surface of the material is often sufficient to provide the necessary information. In other cases, degradation of materials occurs initially at the surface, so that surface analysis is the appropriate route for assessing the state of the material. Techniques for surface analysis or characterization therefore play an important role in materials characterization and NDE. Many of these involve the exposure of the surface to different forms of radiation — photons, x-rays, electrons, neutrons, or ions, and then detecting and analyzing the resulting radiations that are emitted, as depicted in Figure 6.4. A few analytical techniques have developed from basic studies of the interactions of radiation with surfaces, and these have become standardized methods [6]. Typical applications involve detection of surface contamination, chemical changes in the surface and subsurface, and evaluation of the results of surface cleaning. These techniques depend on the electron energy levels of atoms on the surface (Figure 6.5).

6.2.7 X-RAY FLUORESCENCE (XRF)

One widely used method of surface chemical analysis, known as x-ray fluorescence (XRF), exposes the surface to x-rays and the energy of these x-rays excites electrons to higher energy levels between energy states in the atoms at the surface of the material. To explain how this method works, it is necessary to consider the electron energy levels of the atoms in the surface as shown in Figure 6.5a. Electrons from the lowest energy K shell can be excited to higher energy shells such as the L shell, or even the M or N shells. In lighter elements, only the L shell needs be considered but, for heavier elements, M and N shells come under consideration. However, the probability of an M-to-K electron transition is about an order of magnitude less than for an L-to-K transition.

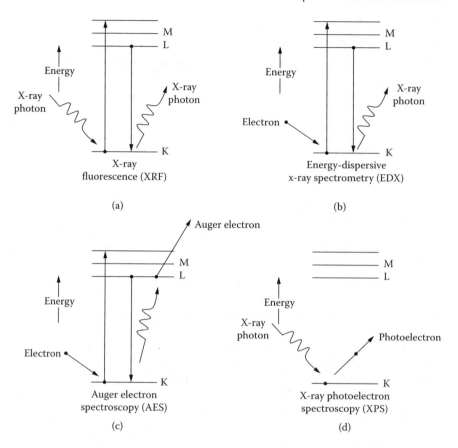

FIGURE 6.5 Electron energy diagrams and mechanism of the electronic processes underlying the related surface analysis techniques (a) XRF, (b) EDS, (c) AES, and (d) XPS showing energy levels for K, L, M, and N shells and associated transitions.

The excited electrons revert to their lower energy state with emission of another x-ray. This second x-ray has an energy equal to the difference in energies of the two electron states and this energy is therefore characteristic of the element. Therefore, the elements present on the surface can be identified from the energies of the emitted x-rays. The depth of analysis is typically a few micrometers, and the surface area examined is typically 10 μm across. The depth is determined by the "escape depth" of the x-ray photons emitted from the sample, and the area is determined by the diameter to which the x-ray beam can be focused.

6.2.8 Energy Dispersive Spectroscopy (EDS)

Another method that is widely used, especially in conjunction with scanning electron microscopy, is energy dispersive spectroscopy (EDS), which is sometimes known as energy dispersive x-ray spectroscopy (EDX). This is often incorporated as a module with scanning electron microscopes. In this case, the surface is exposed to incident electrons that excite the inner shell electrons in the surface atoms to higher energy states. The resulting condition of the atom is unstable and eventually leads to de-excitation of the electron with consequent emission of an x-ray, as shown in Figure 6.5b. Once again, the energy of the emitted x-ray is characteristic of the atoms on the surface. The emission of x-rays in EDS is more pronounced for the heavier elements than the lighter elements and, therefore, this method is practically limited to detection of elements heavier than beryllium. The electron beam can be focused to a much smaller spot size than an x-ray beam, so this method can be used to analyze surface regions with smaller diameters than XRF, typically 1 μm across. The electron beam typically penetrates to a depth of 5 μm, and the escaping x-rays therefore provide information of the surface to this depth. In these cases, if the surface analysis requires information over only a few atomic layers, then the EDS method is not suitable, and the method of Auger electron spectroscopy — which samples a shallower depth — is more appropriate.

6.2.9 AUGER ELECTRON SPECTROSCOPY (AES)

It is also possible to use incident electrons to excite secondary electrons. This method, known as Auger electron spectroscopy (AES) [7], has the advantage of providing information from much shallower depths of the sample, typically a few nanometers, meaning up to ten atomic layers. Therefore this method is better suited than EDS to analyze the top few layers of atoms at the surface. The mechanism involves the creation of an x-ray photon, as in EDS, but in this case, the x-ray photon is reabsorbed when it knocks an electron out of a higher energy level state within an atom, see Figure 6.5c. The emitted electron, known as an "Auger electron" has a substantially smaller escape depth from the surface than the emitted x-ray photon, so that any Auger electrons detected above the surface will have come from a much shallower depth (1–5 nm) than any emitted x-rays. The area from which the Auger electrons are emitted, typically 100 nm across, is also more precisely defined than the area for EDS emissions. AES works well on conductors and semiconductors, but not as well on insulators. It can be used to detect all elements except hydrogen and helium, and it can give results on chemical compositions with precisions of 0.1 to 1.0 atomic% for most elements.

Each element has a characteristic set of electron energy levels that give rise to a unique set of peaks in the Auger spectrum. These are used to identify the chemical species present (Figure 6.6). The probability of Auger emission is greater for lighter elements, therefore this technique works better for these elements, whereas EDS works better for heavy elements. Auger is also more appropriate than EDS for analyzing submicrometer particles, because the analyzed volume in AES is much smaller than in EDS, as shown in Table 6.3.

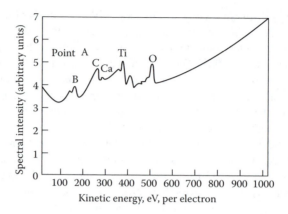

FIGURE 6.6 A typical Auger electron spectrum showing the characteristic variation of intensity vs. electron energy that can be used as a signature for the presence of a particular chemical species.

6.2.10 X-RAY PHOTOELECTRON SPECTROSCOPY (XPS)

It is possible to use x-rays, or in some cases, even UV photons, to excite electrons to sufficiently high energy states that they escape the material. This is the basis for the well-known photoelectric effect, and the analysis technique is known as x-ray photoelectron spectroscopy (XPS) [8]. The energies of the emitted electrons are characteristic of the electron energy levels of the atoms on the surface, Figure 6.5d. The emitted "photoelectrons" can be detected above the surface. The XPS process can be considered the "inverse" of EDS, in which x-rays liberate emitted electrons instead of electrons liberating emitted x-rays.

This phenomenon has been developed into a technique that is widely applicable for extracting information about both elemental and chemical structures at the surface of a material. Similar to AES, the XPS method samples a surface to a depth of a few nanometers, which is determined again by the escape range of the electrons. But the area of analysis, typically 10 μm across, is somewhat larger than in AES because it is not as easy to focus the x-ray beam as the electron beam.

TABLE 6.3
Summary of Characteristics of Various Surface Chemical Analysis Methods

	Incident	Emitted	Typical Spot Size (μm)	Typical Depth
XRF	X-ray photon	X-ray photon	10 μm	1 μm
EDS	Electron	X-ray photon	1 μm	1 μm
AES	Electron	Electron	1 μm	5 nm
XPS	X-ray photon	Electron	10 μm	5 nm

All elements can be detected, except hydrogen and helium. XPS works well on both conducting and insulating materials. The method can give results with a precision up to 0.01 to 0.1 atomic% for most elements.

6.3 EXPOSURE, DOSE, AND DOSE RATE

When a material is subjected to radiation, the exposure ε is the product of both intensity I and time t

$$\varepsilon = I \cdot t. \tag{6.15}$$

The main point is that the exposure is proportional to intensity and time. The units that are used for ε and I can vary, but often ε is expressed in "ion pairs per unit volume," and the unit for this is the Roentgen. This implies that intensity must be expressed in "ion pairs per unit volume per unit time."

6.3.1 REDUCTION IN INTENSITY OF A DIVERGENT BEAM OF RADIATION

If we consider a source radiating in all directions, for example, the radiation emitted by a radioactive source such as Co^{60} or Ir^{192}, both sources of gamma rays, in the absence of shielding, the intensity of radiation I will decay with the square of the radial distance r,

$$I(r) = \frac{I_o}{4\pi r^2}. \tag{6.16}$$

6.3.2 SHIELDING OF A NONDIVERGENT BEAM OF RADIATION

The intensity of radiation, and hence the exposure, can be reduced by shielding, which refers to placing an absorbing material in the path of the beam of radiation. For example, if a nondivergent beam of radiation passes through a material, the intensity of radiation decays exponentially with distance. Using the same equation as is used for attenuation,

$$I(x) = I(0) \exp(-\mu x), \tag{6.17}$$

with the rate of decay depending on the attenuation coefficient, μ, of the particular material. The exposure, therefore, also decreases exponentially with the thickness of the shielding material,

$$\varepsilon(x, t) = I(x)t = I(0) \exp(-\mu x)t. \tag{6.18}$$

6.3.3 REDUCTION IN INTENSITY OF A DIVERGENT
BEAM WITH SHIELDING

With both divergence and shielding, the intensity of radiation decays with both radial distance from the source and thickness of the shield and, therefore an equation that combines both the exponential and the $1/r^2$ decrease in intensity is needed,

$$I(r, x) = \frac{I_o}{4\pi r^2} \exp(-\mu x). \tag{6.19}$$

6.3.4 DOSE

Exposure to radiation can have biological effects, many of which are undesirable, for example, in the case of human health. However, it should be noted that there are some therapeutic uses of high-intensity radiation, and some diagnostic uses of low-intensity radiation which are intended to be nondestructive.

The biological effect of exposure to radiation is known as the "dose," D. For a given type of radiation the dose is proportional to the exposure, and, therefore for a nondivergent beam of radiation,

$$D(x, t) = D_R(0) \cdot \exp(-\mu x) \cdot t, \tag{6.20}$$

where $D_R(0)$ is the dose rate in the absence of shielding. The dose rate $D_R(x)$ as a result of shielding of thickness x can be expressed as

$$D_R(x) = D_R(0) \cdot \exp(-\mu x). \tag{6.21}$$

This means that the exposure, dose rate, and dose can be reduced by shielding using a material with high attenuation coefficient μ, and/or high thickness x.

6.3.5 REDUCTION OF EXPOSURE TO RADIATION AND DOSE RATE

The exposure to radiation can be reduced either by reducing the intensity I of the radiation, or by reducing the time t spent in the radiation. The former can be reduced either by being far from the source, in the case of a divergent beam of radiation (increasing the radial distance r), or by using shielding materials with high attenuation coefficient μ or thickness x to absorb or attenuate the radiation intensity. In general, we can state that the dose rate, which is proportional to the intensity of radiation, can be expressed by an equation of the form

$$D_R(x, r) = D_R(0, 1) \cdot \frac{1}{r^2} \exp(-\mu x), \tag{6.22}$$

where $D_R(0, 1)$ is the dose rate at unit radial distance ($r = 1$) from a divergent source of radiation when there is no shielding material ($x = 0$).

The total absorbed dose $D(x, r, t)$, which is given by the following equation,

$$D(x, r, t) = D(0, 1) \cdot \frac{1}{r^2} \exp(-\mu x) \cdot t \tag{6.23}$$

can be limited by minimizing the time of exposure t.

6.3.6 MEASUREMENT UNITS: ROENTGEN, RAD, AND REM

Exposure, which is the amount of ionization per unit mass, is measured in Roentgens. Intensity, which is the rate of ion generation, is measured in Roentgens per second. The absorbed dose, which is the amount of energy deposited per unit mass, is measured in rads, and the absorbed dose rate is therefore in rads per second. The biologically effective dose is measured in rems (which stands for "Roentgen equivalent man"). The biologically effective dose rate is then measured in rems per second [5, Ch. 5].

> *Roentgen:* This is a measure of the amount of ionization. By definition the Roentgen is: *An exposure that produces 2.58×10^{-4} coulombs of charge per kilogram of air (0.333×10^{-3}) coulombs per cubic metre)*
>
> *Rad:* This is a measure of the absorbed dose. By definition the rad is: *An energy deposition of 0.01 joules per kilogram.* 0.869 Rad in air is equivalent to 1 Roentgen. The SI unit of absorbed does is the Gray (Gy), which is *an energy deposition of 1 joule per kilogram.* Therefore 1 rad = 0.01 Gy.
>
> *Rem:* This is a measure of the biologically effective dose, which is a weighted version of the amount of energy deposited per unit mass, taking into account that certain types of radiation have greater biological effects than others. The Rem is equal to the number of Rads multiplied by the relative biological effectiveness for the radiation. The SI unit of biologically effective does is the Sievert (Sv), which is the product of the absorbed does in Grays and the RBE for the particular type of radiation. Therefore 1 Sv = 100 rem. The relative biological effectiveness (RBE) is 1 for x-rays, gamma rays, and beta rays, with energies > 30keV, 1.7 for beta rays with energies < 30keV, 3, 4 or 5 for thermal neutrons, 10 for fast neutrons, protons and alpha particles and 20 for heavy ions.

WORKED EXAMPLE

In one year, an NDE worker received a dose of 2 rad of gamma rays, 0.5 rad of thermal neutrons, and 0.1 rad of fast neutrons. What is the total biologically effective dose?

Solution

Radiation	Quality Factor (QF) (rads)	Absorbed Dose	Biologically Effective Dose (rems)
Gamma	1	2	2
Thermal neutrons	3	0.5	1.5
Fast neutrons	10	0.1	1
Total		2.6	4.5

6.3.7 RECOMMENDED UPPER LIMITS FOR RADIATION DOSE

A designated radiation worker should have a whole body dose of less than 5 rem/year and should not exceed an accumulated dose of $5 \times (N\text{-}18)$ rem, where N is the age of the worker in years.

For a member of the general public, the whole body dose is an order of magnitude less than this, so that the whole body dose should be less than 0.5 rem/year and should not exceed $0.5 \times (N\text{-}18)$ rem accumulated dose.

EXERCISES: EFFECTS OF RADIATION ON MATERIALS

6.1 A 2.25 Ci cobalt-60 gamma radiation source is to be used with a scintillation counter as detector to examine a 50 mm thick lead container. What percentage change in count rate will a 3 mm deep cavity in lead cause, if the half-value thickness of lead for this radiation is 12 mm?

6.2 A thin layer comprised of 50% nickel and 50% titanium needs to be inspected for segregation effects (inhomogeneities). Using the information in Table 6.1, select a suitable wavelength of x-rays that can best distinguish between the Ni and Ti regions and calculate (1) the expected linear attenuation coefficient for the material assuming it is homogeneous, (2) the linear attenuation coefficient for x-rays passing through nickel, and (3) the linear attenuation coefficient for x-rays passing through titanium. Calculate the ratio of intensities of transmitted radiation between the Ni and Ti regions, and between the Ti and homogenous Ni-50%Ti alloy regions, if the specimen is 0.1 mm thick. (densities: for Ni, $\rho = 8850$ kg/m³ and for Ti, $\rho = 4540$ kg/m³).

6.3 A source of radiation produces a biological dose rate of 200 rem per year at a distance of 1 m when unshielded. What is the unshielded dose rate at a distance of 5 m and what thickness of lead shielding is needed to reduce that dose rate to below 0.5 rem per year at a working distance of 5 m, assuming that lead has a half-value thickness of 5 mm for the type of radiation under consideration.

REFERENCES

1. Casanova, E.J., Some important basics in radiography, *Mater Eval* 44, 151, February 1986.
2. Kriesz, H., Radiographic NDT — a review, *NDT Int* 12, 270, 1979.
3. Iddings, F.A., Radiation detection for radiography, *Mater Eval* 59, 926, August 2001.
4. Halmshaw, R., Industrial radiology: theory and practice, Chapman and Hall, London 1995.
5. Becker, G.L., *Radiographic NDT*, DuPont, Wilmington, 1990.
6. Newman, J.G., Chemical characterization of surfaces, *ASM Handbook: Failure Analysis and Prevention*, Vol. 11, ASM, Materials Park, OH, 2002. p. 527.
7. Childs, K., *Handbook of Auger Electron Spectroscopy*, Chastain, J. and King, R.C., Jr. (Eds.), Physical Electronics, MN, 1995.
8. Moulder, J., *Handbook of X Ray Photoelectron Spectroscopy*, Chastain, J. and King, R.C., Jr. (Eds.), Physical Electronics, MN 1995.

FURTHER READING

Becker, G.L., *Radiographic NDT*, DuPont, Wilmington, 1990.
Doo, B., Basic theory and practice of radiography, *NDT Int* 2, 161, 1969.
Halmshaw, R., *The Physics of Industrial Radiology*, Elsevier, New York, 1966.
Halmshaw, R., *Industrial Radiology: Theory and Practice*, Chapman and Hall, London, 1995.
http://www.ndt-ed.org/EducationResources/CommunityCollege/Radiography/cc_rad_index.htm.
Johns, H.R. and Cunningham, J.R., *The Physics of Radiology*, 4th ed., Thomas Publishers, Springfield, IL, 1983.

7 Mechanical Testing Methods

In this chapter, we consider the various tests for determining the mechanical properties of materials. These include tensile tests using application of load on a specimen and measurement of the resulting deformation, hardness tests using surface indentation methods, crack tests using liquid penetrants and other related methods, and fracture tests using impact methods. Some of these tests are destructive and some are nondestructive.

7.1 TENSILE TESTING

Tensile tests usually apply controlled loads to samples and measure the deformation (strain) of the sample. The stress can be determined exactly by knowing the cross-sectional area of the sample, known as the true stress, but more often the approximate stress is calculated assuming that the cross-sectional area remains the same throughout. This is known as the engineering stress.

Tensile tests are used to provide mechanical property data of materials. The tests are nondestructive up to the yield point. In this region, the stress–strain curve shows a linear relationship, and the slope of the graphs of stress against strain is the elastic modulus. Beyond the yield point, the material suffers permanent plastic deformation, and at the ultimate tensile strength the material will fail.

The equipment used for tensile tests may be screw- or hydraulically-driven. An example of a screw-driven system is shown in Figure 7.1.

7.1.1 STRESS–STRAIN CURVE

A typical stress–strain curve that would result from such a tensile test is shown in Figure 7.2. Several mechanical properties of a material can be obtained from a stress–strain curve. These include elastic modulus, yield strength, ultimate tensile strength, ductility, resilience, and toughness [1]. From Figure 7.2, it can be seen that at low stresses, below the yield strength, the variation of strain with stress is linear, meaning that the elastic modulus remains constant. Because it is difficult to tell exactly where the stress–strain curve becomes nonlinear, the *yield strength* is, by convention, defined as the stress at which the residual plastic deformation is 0.2%.

7.1.2 ENGINEERING STRESS VS. TRUE STRESS

Beyond the yield strength, the strain increases more rapidly with stress, and the stress reaches a peak known as the ultimate tensile strength. On a plot of true stress

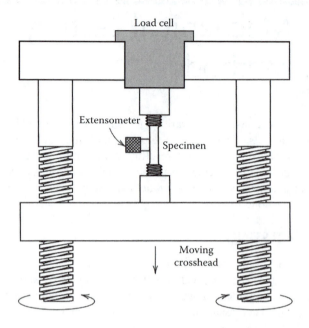

FIGURE 7.1 Schematic of a screw-driven apparatus for carrying out tensile tests on materials.

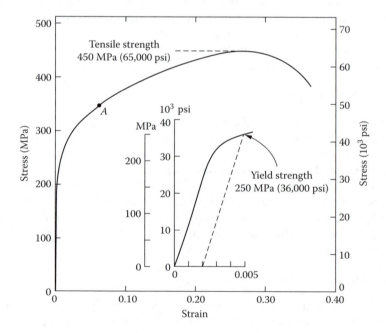

FIGURE 7.2 The stress–strain curve obtained from a tensile test.

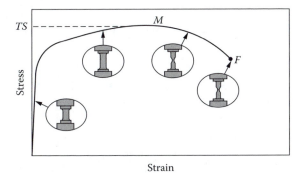

FIGURE 7.3 Typical variation of stress with strain for a metal.

vs. strain, the graph continues to show a monotonic increase of stress with strain. On a plot of engineering stress vs. strain, the apparent stress needed to produce a given strain starts to decrease beyond a certain value of stress due to the reduction in area of the sample (Figure 7.3).

7.1.3 NONLINEAR BEHAVIOR

The linear region of the stress–strain curve can be described by Hooke's law, as discussed in Chapter 1. However, a different form of equation is needed to describe the nonlinear regions of the stress–strain curve. This form of equation is

$$\sigma_t = K\varepsilon_t^n, \tag{7.1}$$

where σ_t is the true stress, ε is the strain, and K and n are constants. This equation can be used to fit a very wide range of stress–strain curves in terms of only two parameters. This can help to easily quantify stress–strain curves in terms of only two descriptors, n the strain-hardening exponent, and K the strength coefficient. K clearly has similarities to the elastic modulus, whereas the strain exponent n describes the curvature.

WORKED EXAMPLE

Determine the strain-hardening exponent for an alloy with true stress of 415 MPa and true strain of 0.1, with $K = 1035$ MPa.

Solution

$$\sigma_t = K\varepsilon_t^n, \tag{7.2}$$

$$\log e\left(\frac{\sigma_t}{K}\right) = n\log\varepsilon_t \tag{7.3}$$

$$n = \frac{\log_e \left(\dfrac{\sigma_{true}}{K} \right)}{\log_e \varepsilon} \tag{7.4}$$

$$n = \frac{\log_e \sigma_t - \log_e K}{\log_e \varepsilon_t} \tag{7.5}$$

$$n = \frac{19.84 - 20.76}{-2.30} = 0.4 \tag{7.6}$$

It has been shown [2] that mechanical properties can be measured indirectly through utilization of correlations with other properties of materials, such as ultrasonic, magnetic, or electrical properties. For example, the mechanical properties of materials are closely related to the ultrasonic velocity and ultrasonic modulus [3], so that properties like flexural strength and ultimate tensile strength can be determined indirectly from ultrasonic measurements. In many cases, these indirect measurements provide the only available practical nondestructive test methods, because traditional mechanical testing on the parts is not possible.

7.2 HARDNESS TESTS

Hardness is the resistance of a material to plastic deformation. It is usually measured by an indentation test, which consists of applying a known load over a particular geometry and seeing how far the material deforms. There are various forms of the test, ranging from the original Brinell test, which uses a spherical indenter ("ball indenter"), through a variety of cone-shaped indenters used in some of the Rockwell tests, to pyramid-shaped indenters used in the Vickers and Knoop indentation tests. In all cases, the deformation is measured as a result of application of load under standardized conditions.

All the indentation tests induce some permanent plastic deformation on the surface of the material under test. This can be thought of as a localized plastic zone. The area of damage varies significantly depending on the type of test. Consequently, these tests are not truly nondestructive. Also, it is known that the resulting calculated hardness varies with load, which can be a problem; so it is best if all hardness tests are conducted under identical standardized conditions.

7.2.1 COMPARISON OF HARDNESS TESTS AND CONVERSION BETWEEN HARDNESS SCALES

The four main hardness tests have various relative advantages and disadvantages.

The Brinell test, which was the first standard hardness test, is hardly used today. It gives a good average value of hardness for the material under test because of the large indentation. However, for the same reason this is not suitable for

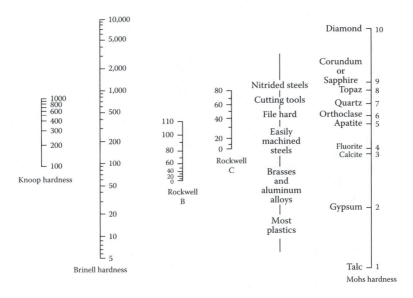

FIGURE 7.4 Comparison of hardness scales for the Brinell, Rockwell, and Vickers hardness tests.

small specimens. The Knoop test is good for brittle materials, and it results in the smallest indentation among the tests that are being considered here. It does, however, require surface preparation. The Rockwell test is very simple and requires no surface preparation; the test is widely used on metals. The Vickers test can be used on all materials. However, it requires also surface preparation, such as grinding and polishing of the surface to get meaningful results.

The values of hardness obtained from the various tests can be compared, and conversions can be made from one scale to another. A conversion chart is shown in Figure 7.4.

7.2.2 RELATIONSHIP OF HARDNESS TO OTHER MECHANICAL PROPERTIES

As discussed earlier, the dimensions of the hardness are force per unit area, the same dimensions as stress. Because the hardness of a material is a measure of its resistance to plastic deformation, it is reasonable to expect some relationship between hardness and other mechanical properties related to permanent deformation, such as yield strength or ultimate tensile strength. For example, the correlations between the Brinell hardness and both yield strength and tensile strength of ductile iron are shown in Figure 7.5. From this, it appears that a calibration curve can be constructed over the range BHN 160–300, which is very close to linear for tensile strengths over the range 400–1100 MPa, and for yield strengths over the range 300–900 MPa.

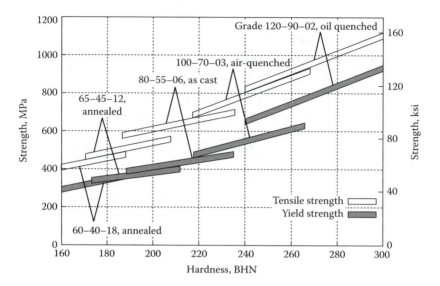

FIGURE 7.5 Variation of yield strength and tensile strength with hardness for ductile iron showing that hardness is closely related to the strength of a material.

Some authors give equations that relate yield strength and ultimate tensile strength with hardness. For example, Meyers and Chawla [4] state that yield strength, σ_y, is equal to one third of the ratio of load over projected area of indentation in a hardness test

$$\sigma_y = \frac{1}{3}\frac{F}{A_P}, \tag{7.7}$$

where F is the applied load that produces the indentation, and A_p is the projected area of the indentation.

An empirical relationship between tensile strength in MPa and Brinell hardness has been suggested,

$$\sigma_{uts} = 3.45 \text{ BHN}.$$

An approximate relationship also exists between the fatigue limit, σ_{fl}, and the ultimate tensile strength, σ_{uts}, of a material,

$$\sigma_{fl} = 0.5\,\sigma_{uts}.$$

Therefore, the results of a hardness test should give some indication of the stress amplitude above which fatigue failure will occur.

Determine the fatigue limit for a material with a Brinell indentation, under standard conditions, of 2.74 mm.

Solution

$$HB = \frac{2F}{\pi D(D - \sqrt{D^2 - d^2})} = 500 \tag{7.8}$$

$$\sigma_{uts} = 3.45 \text{ BHN} = 1725 \text{ MPa} \tag{7.9}$$

$$\sigma_{fl} = 0.5 \, \sigma_{uts} = 862 \text{ MPa} \tag{7.10}$$

7.2.3 VARIATION OF INDENTATION AREA WITH LOAD: RELIABILITY OF HARDNESS TESTS

At low loads, the measurement of hardness can result in significant errors. As an example, the variation with load of both Knoop and Vickers hardness is shown in Figure 7.6. It can be seen that above a load of 100 g the hardness values are stable, but below 100-g loads, and particularly at loads below 25 g, the measured values change markedly.

These variations in measured microhardness arise because the area of the indentation does not vary linearly with load in the low-load regime, as shown in Figure 7.7.

In these cases, the surface condition, particularly if it has been improperly prepared, can adversely affect the measurement of hardness. Electropolishing of

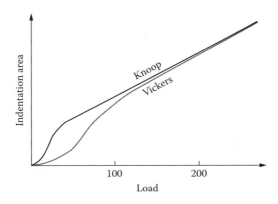

FIGURE 7.6 Variation of indentation area with load for both Knoop and Vickers hardness measurements. A linear relationship yields a constant value of hardness.

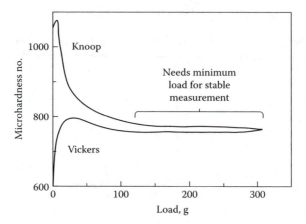

FIGURE 7.7 Variation of microhardness with load in the low-load regime.

the surfaces is desirable, although not absolutely essential for these low-load measurements. It can therefore be concluded that in practical terms, a minimum load is needed for stable measurements. Once beyond such minimum load, and therefore into the stable regime, the area of indentation becomes a linear function of load, as shown in Figure 7.8.

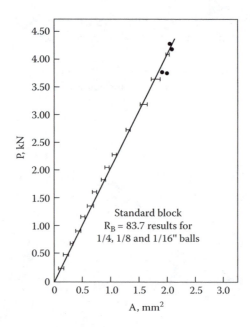

FIGURE 7.8 Variation of indentation area with load for Rockwell B scale hardness measurements.

7.3 CRACKS AND FAILURE OF MATERIALS

When materials fail, they do so by the propagation of a crack or cracks. Often, it is necessary to know the size of a crack, because the critical stress needed to cause a crack to grow decreases with the size of the crack. The simplest nondestructive evaluation method for crack-sizing is the use of liquid penetrants. These are colored liquids that are sprayed on to the surface of a material and are used to highlight the presence of cracks.

The propagation of cracks is shown in Figure 7.9. The region close to the crack tip is a region of high stress, so this is where the crack grows. Once the tip of the crack has been passed, the stress level reduces, and further cracking occurs at the new location of the crack tip.

The presence of cracks always precedes failure in materials. Examples are the generation of cracks in a material under the action of a unidirectional load that can lead to failure of the material at the ultimate tensile strength; and the failure of a material under the action of a periodically varying stress, a process known as *fatigue*, in which the material fails at stress levels below the tensile strength. Figure 7.10 shows the fracture surface of a material that failed through fatigue. It can be seen that the crack grew in a series of stages, which left characteristic features known as beachmarks that mark the progress of the crack over time.

7.3.1 CRACKS AND OTHER DEFECTS

There are a number of nondestructive ways to detect cracks in materials. These methods include ultrasonics in which sound waves are reflected from the surfaces of cracks; radiography, in which the attenuation of x-rays is reduced in the

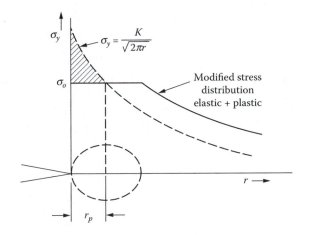

FIGURE 7.9 Stress profile in the vicinity of a crack.

FIGURE 7.10 Beachmarks on the fracture surface are a sign of fatigue failure.

vicinity of cracks, and therefore a map of the transmitted x-ray intensity shows the location of cracks; eddy currents, in which the passage of alternating electrical currents in the material is disturbed by the presence of cracks, and these disturbances are detected as a change in impedance of a coil; liquid penetrants, in which the size and location of a crack is indicated from the amount of liquid that is absorbed by the crack; and magnetic particle inspection, in which the presence of a crack in the material causes disruption of the magnetic flux lines, which then attract small magnetic particles to the location of the crack when the material is magnetized. These methods can all be used to detect both surface-breaking and subsurface flaws, and each will be dealt with in later chapters.

Mechanical impedance measurements have been used [5] to detect the presence of near-surface defects in plates and panels. From these tests, it was found that the mechanical impedance method had some advantages over traditional ultrasonics and could be used where no couplant was needed between the transducer and test material. Surface scans were used to produce images, which were defect maps of the surface. However, the method was not so useful for small defects. Disbonds of 10-mm diameter were detectable using this method, with the actual limit of detectability being dependent on the thickness of the sample and the depth of the defect below the surface.

When a material is under stress and has cracks, it may or may not fail, depending on the level of stress and the size of the cracks. Materials that are under stress and have cracks can fail when the temperature changes, even though the stress and crack sizes have remained the same, because the resistance of the material to crack growth can be impaired as the temperature decreases. This is the problem of ductile-to-brittle transition. An interesting example is shown in

FIGURE 7.11 Oil tanker broken in half by crack propagation.

the Figure 7.11, in which the ship that had cracks and was under stress, was able to sustain the stress at higher temperatures, but as the temperature dropped, the fracture toughness was reduced and the critical crack size was thereby reduced until it became the same as the size of the actual cracks in the material. Once that happened, the cracks began to grow within the material without further application of stress.

7.3.2 CRACK GROWTH

Under given conditions of stress and temperature in a particular material, the stress needed to cause a crack to continue to grow decreases with the length of the crack. The relationship between the crack size and stress needed to cause the crack to grow is given by the fracture toughness, which is discussed below. This also means that under given conditions of stress and temperature, there is a critical crack size beyond which any crack will just continue to grow, without further application of stress, resulting in failure. Determination of the presence of cracks at or above the critical crack size is then a vital component of nondestructive evaluation of materials.

7.3.3 FATIGUE

It is important to remember that materials can fail when subjected to repeated cyclic stress at levels below the ultimate tensile strength, and in some cases, at levels below the yield strength. This means that although a material has been able to survive exposure to the same stress level many times before, it can still fail because of an accumulation of damage.

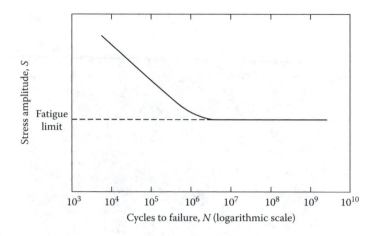

FIGURE 7.12 Number of cycles to failure decreases with stress/strain amplitude.

This phenomenon of failure after a large number of stress cycles is usually represented on a graph of number of cycles needed to reach failure against the stress amplitude, as shown Figure 7.12. Note that the number of stress cycles to reach failure decreases with increasing stress/strain amplitude, beginning with one stress cycle at the tensile strength. In some materials, there is a threshold value of stress amplitude below which failure does not seem to occur; in other words, the stress amplitude at which the number of cycles needed to cause failure suddenly becomes infinite. This threshold level of stress is known as the *endurance limit*. In some other materials there does not appear to be an endurance limit, so failure continues to occur even at the smallest stress amplitudes, although the number of cycles to failure increases as the stress amplitude decreases, as shown in Figure 7.13.

7.3.4 DETECTION OF CRACKS USING LIQUID PENETRANTS

If the flaws are surface-breaking, a simple method for detecting their presence, and even to some extent sizing the flaws, is liquid penetrant inspection [6,7]. This offers significant improvements over unaided visual inspection in terms of detectability of small cracks [8]. The method is largely qualitative, although the extent of the flaws across the surface can be measured. The depth of the flaw is however much more difficult to determine by this method.

Liquid penetrant inspection is used primarily with nonmagnetic materials, because magnetic particle inspection (MPI) is used for similar purposes on magnetic materials. In most cases, the liquid is fluorescent, which means that the presence of the liquid on the surface can more easily be seen, particularly under ultraviolet light, which thereby causes the presence of the penetrant to stand out against the background. The fluorescent penetrant method is often referred to as "FPI."

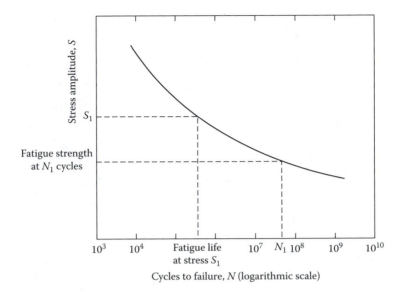

FIGURE 7.13 Failure occurs at stress below yield strength.

The normal procedure is to clean the surface of the test material, apply a coat of liquid penetrant, allow the coating to dry, and then clean the surface using a dry method to remove excess penetrant located solely on the surface. The liquid penetrants can seep into cracks as small as 1 μm by capillary action. Once the surface has been cleaned, a coating of developer is applied, which causes the penetrant located within cracks to seep out, thereby revealing the location of cracks [9]. A flowchart summarizing the procedure is shown in Figure 7.14. To some extent, it can be argued that the amount of penetrant seeping out is proportional to the volume of the crack, so that there should be a relationship between this and the flaw size, but in practice this is not so simple. Using this method, flaws as small as 130 nm (5 μ in.) in length can be detected [10].

The penetrant enters the cracks when the liquid coating is applied, as shown in Figure 7.15a, and remains in the cracks after the surface has been cleaned, as shown in Figure 7.15b; however, particularly for deep narrow cracks, the indication of the existence of the crack may only be revealed when the developer is applied (Figure 7.15c). Specifications and standards for penetrant inspection are being improved and effectively being made more restrictive. Despite its apparent simplicity, there are detailed inspection procedures and recommendations for the use of this method that have been drawn up [11]. Concerns over the use of certain organic penetrants have also led to tighter regulations [12].

The detection of small surface-breaking cracks using liquid penetrants can be aided by the use of appropriate illumination. In particular, the use of fluorescent liquids in combination with ultraviolet lighting can greatly enhance the brightness and contrast of the reflected signal from the small areas of penetrant

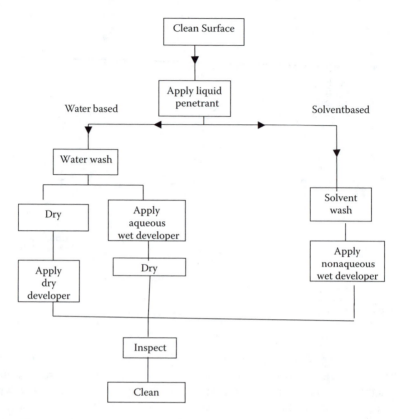

FIGURE 7.14 Flowchart showing the basic steps in preparation of a material for liquid penetrant inspection.

on the surface relative to the background, in this case the bare material. Normal light illumination in the visible range of the spectrum does help somewhat, but this also increases the reflected signal from the background. In the case of ultraviolet light, however, the normal reflectance from the surface in the ultraviolet range will not be visible to the unaided eye. The light that is reflected from the regions with fluorescent liquid penetrant undergoes an energy conversion, which results in the reflection of a wavelength characteristic of the liquid penetrant in the visible part of the spectrum. This results in a beneficial enhancement of the signals from the crack compared with the signals from the integral regions of the material. More recently, laser-based illumination for liquid penetrant inspection has been developed with a consequent improvement in the detectability of small cracks and even some subsurface defects [13].

The appearance of the cracks as a result of liquid penetrant inspection can vary greatly. Three simple examples are given in Figure 7.16.

The use of fluorescent penetrants for crack detection has developed into a method that is finding widespread practical applications because of several

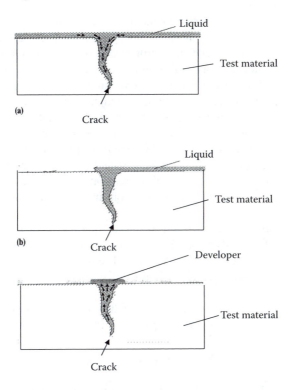

FIGURE 7.15 Three stages in identifying a surface-breaking crack using liquid penetrants.

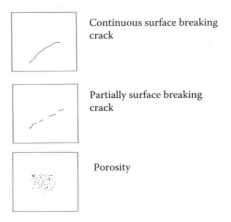

FIGURE 7.16 Examples of the appearance of the surface after liquid penetration: (a) A continuous surface-breaking crack; (b) A partially surface-breaking crack; and, (c) surface-breaking porosity.

advantages that it confers over other methods. The method works on nonmagnetic and magnetic materials (although mostly used on nonmagnetic materials), it is low cost and portable, and the results are easily interpreted. It also allows whole field inspection for parts of complex geometries. FPI is the most widespread method for nondestructive inspection of aircraft engine components, with over 90% of these being inspected by this method during their lifetime [14].

There are also some limitations to the method. Specifically, the discontinuities must be surface-breaking for the method to work; test samples must be thoroughly cleaned before the test because surface films or dirt can prevent the liquid penetrating the cracks, leading to poor inspection results; and penetrants can be a source of surface contamination that interferes with subsequent surface chemical analysis. The method is not well-suited to inspection of porous materials, such as powder metallurgical parts [15].

Welds are regions of engineering materials that are particularly susceptible to failure because of local variations in microstructure, variations in mechanical properties, and high levels of residual stress. The detection of cracks in welds is of great interest. The surfaces of materials in the vicinity of welds are generally not very smooth because of the weld crown and, to a lesser extent, the variation in structure in the heat-affected zones, so that the use of liquid penetrants in welds presents greater problems than in the parent material.

7.3.5 OTHER METHODS FOR SURFACE INSPECTION

More advanced methods of nondestructive evaluation of mechanical properties using optical methods include image processing techniques. The test material can be excited mechanically by an acoustic stressing mechanism, and using the fact that the regions with defects have different mechanical properties from the rest of the material [16], an image can be produced. Addition/subtraction of optical fringe patterns under stress can then be used to produce a shearograph pattern that can be enhanced by image processing using digital signal processing methods.

7.4 IMPACT AND FRACTURE TESTS

Impact tests measure how much energy is needed to cause fracture of a material under standardized conditions of sample shape, dimension, and with a known groove machined into the sample. From these tests, the impact energy for the given geometry can be found, and this is directly related to the fracture toughness of the material, which defines the relationship between critical stress needed to cause a crack to grow and the length of the crack.

To test the resistance of a material to crack growth, the standard method is the Charpy impact test, which is shown in Figure 7.17. This is unquestionably a destructive test, because the objective is to find out how much energy is needed to break the sample. However, the results can be used as a calibration procedure for alternative nondestructive tests, by measuring other nondestructive properties of

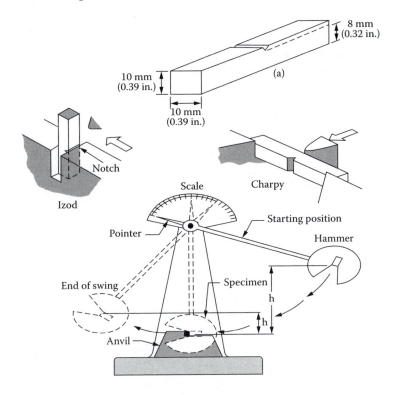

FIGURE 7.17 Details of the Charpy impact test for determination of fracture toughness of materials.

the material and correlating these with the results of the Charpy test or the calculated fracture toughness.

7.4.1 FRACTURE TOUGHNESS

The relationship between the crack size and stress needed to cause the crack to grow is known as the fracture toughness, usually denoted by K_{1C}. So, under somewhat idealized conditions known as plane strain conditions, if there is a crack of length a, the critical stress σ_c that is needed to cause the crack to grow is related to the size of the crack and the fracture toughness by the relationship

$$\sigma_c = \frac{K_{1C}}{\sqrt{\pi a}}. \tag{7.11}$$

Table 7.1 shows the yield strength and fracture toughness for various materials.

TABLE 7.1

Room-Temperature Yield Strength and Plane Strain Fracture Toughness Data for Selected Engineering Materials

Material	Yield Strength		K_{Ic}	
	MPa	ksi	$MPa\sqrt{m}$	$ksi\sqrt{in.}$
Metals				
Aluminum Alloy[a] (7075-T651)	495	72	24	22
Aluminum Alloy[a] (2024-T3)	345	50	44	40
Titanium Alloy[a] (Ti-6Al-4V)	830	120	55	50
Alloy Steel[a] (4340 tempered @ 260°C)	1640	238	50.0	45.8
Alloy Steel[a] (4340 tempered @ 425°C)	1420	206	87.4	80.0
Ceramics				
Concrete	—	—	0.2–1.4	0.18–1.27
Soda-Lime Glass	—	—	0.7–0.8	0.64–0.73
Aluminium Oxide	—	—	2.7–4.2	2.5–3.8
Polymers				
Polystyrene (PS)	—	—	0.7–1.1	0.64–1.0
Polymethyl Methacrylate (PMMA)	53.8–73.1	7.8–10.6	0.7–1.6	0.64–1.5
Polycarbonate (PC)	62.1	9.0	2.2	2.0

[a]*Source:* Reprinted with permission, *Advanced Materials and Processes*, ASM Internationl, © 1990.

WORKED EXAMPLE

An ultrasonic NDE technique can detect flaws and cracks greater than 3 mm in 4340 steel. If a material has a fracture toughness of $K_{1C} = 60.4$ MPa.m$^{0.5}$ and a yield strength of 1515 MPa, and is subjected to a stress of half the yield strength, can the flaws be detected before it fails?

Solution

For a given stress, the critical crack size can be defined as

$$a_c = \frac{K_{1C}^2}{\pi\sigma^2} \qquad (7.12)$$

$$a_c = \frac{1}{\pi}\left(\frac{60.4}{707.5}\right)^2 = 2.3\,mm \qquad (7.13)$$

Conclusion: The material fails before the cracks are detectable.

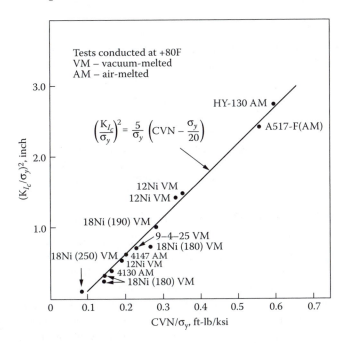

FIGURE 7.18 Graph showing the relationship between fracture toughness, K_{1C}, and Charpy V notch impact energy, CVN. The relationship between CVN and fracture toughness K_{1C} can be approximated by the equation $K_{1C}^2 = \sigma_y^2 (\alpha \frac{CVN}{\sigma_y} - \beta) = \sigma_y \alpha\, CVN - \beta \sigma_y^2$ in which $\alpha = 0.625 \times 10^6\ \text{m}^{-2}$ and $\beta = 0.00626\ \text{m}$.

7.4.2 RELATIONSHIP BETWEEN FRACTURE TOUGHNESS AND CHARPY V NOTCH TEST

The Charpy impact energy has a direct relationship to the fracture toughness. This is shown in Figure 7.18, which is a plot of the square of the fracture toughness, K_{1C}, against the Charpy V notch (CVN) impact energy needed to cause failure. An empirical formula is shown, which can be used to represent the linear dependence of K_{1C}^2 on CVN. This simple dependence of fracture toughness on CVN means that we can estimate the fracture toughness from the impact energy obtained from the Charpy test, and this makes the Charpy test a convenient practical method for determination of fracture toughness of materials.

WORKED EXAMPLE

A steel specimen has a 0.2% yield strength of 825 MPa. Its upper shelf energy, as determined from a Charpy test, is 67.8 N.m. Calculate the fracture toughness, K_{1C}.

Solution

Using the equation that relates fracture toughness, K_{IC}, to Charpy V notch impact energy, CVN,

$$K_{IC}^2 = \sigma_y^2 \left(\alpha \frac{CVN}{\sigma_y} - \beta \right) = \sigma_y \alpha CVN - \beta \sigma_y^2 \tag{7.14}$$

$$K_{IC}^2 = 3.62 \times 10^{16} - 0.45 \times 10^{16} = 3.07 \times 10^{16} \tag{7.15}$$

$$K_{IC} = 175 \text{ MPa·m}^{0.5} \tag{7.16}$$

7.4.3 TEMPERATURE DEPENDENCE OF FRACTURE TOUGHNESS

The fracture toughness of a material varies with temperature and this can have rather serious consequences, as the preceding example of the ship clearly demonstrates. The fracture toughness, or equivalently the impact energy, decreases as the temperature decreases, as shown in Figure 7.19 and Figure 7.20. The sharpness of the decrease with temperature is different from material to material, and in some cases, this change in properties with temperature can be sharp enough that the behavior can be described in terms of a ductile-to-brittle transition temperature. This means that at high temperature with a high fracture toughness or high impact energy (upper shelf energy) the material is ductile, but at lower temperature with a low fracture toughness or impact energy (lower shelf energy) the material is brittle.

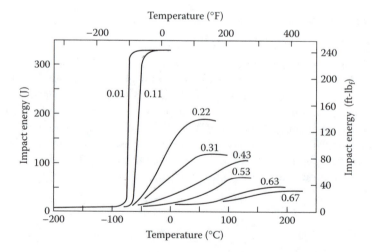

FIGURE 7.19 Variation of the Charpy V notch impact energy with temperature, showing a transition temperature between the high-temperature ductile regime with its upper shelf energy, and the low-temperature brittle regime with its lower shelf energy.

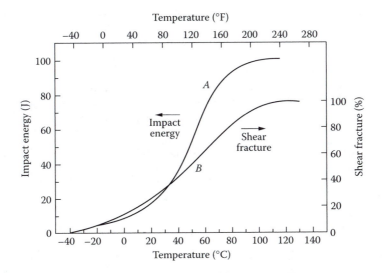

FIGURE 7.20 Variation of the impact energy and shear fracture with temperature, showing a transition temperature for a high-temperature ductile regime and a low-temperature brittle regime.

It should also be noted that the ductile-to-brittle transition temperature can change depending on the treatment of the material or its exposure to different service conditions. A well-known example is the effect of radiation, which causes embrittlement of constructional materials used in nuclear reactors, in which exposure to high levels of radiation over many years causes changes in the structure of the material with an increase in the ductile-to-brittle transition temperature. This can mean that initially, under normal service conditions of temperature and pressure, the constructional material is operating in the ductile regime (upper shelf energy in Figures 7.19 and 7.20), but later as the material degrades, the material may be operating in the brittle regime (lower shelf energy), even though the service conditions of temperature and pressure may be the same as before. This can result in an insidious problem, whereby a material suddenly fails under operating conditions that are identical to conditions under which it has operated safely for perhaps many years.

EXERCISES: MECHANICAL TESTING

7.1 Mechanical measurements have been made on a piece of heat-treated 1020 steel as part of a materials selection search. The data in Table 7.2 were obtained. Determine the strain-hardening exponent and the strength coefficient for this particular heat treatment of 1020 steel.

7.2 A load-bearing component is to be made from 2.25Cr–Mo steel, which has the fracture characteristics shown in Figure 7.21. The stress on the component

TABLE 7.2

	Test Data			Calculated Values		
Engr. Strain ε	Load P kN	Diameter d mm	Engineering Stress σ MPa	True Strain[5] ε	Raw True Stress[5] σ MPa	Corrected True Stress σ_B
0	0	9.11	0	0	0	0
0.0015[1]	19.13	—	293	0.00150	293	—
0.0033[2]	17.21	—	264	0.00329	265	—
0.0050	17.53	—	269	0.00499	270	—
0.0070	17.44	—	268	0.00698	269	—
0.010	17.21	—	264	0.00995	267	—
0.049	20.77	8.89	319	0.0489	335	335
0.218	25.71	8.26	394	0.196	480	461
0.234[3]	25.75	—	395	0.210	488	466
0.306	25.04	7.62	384	0.357	549	501
0.330	23.49	6.99	360	0.530	612	539
0.348	21.35	6.35	328	0.722	674	577
0.360	18.90	5.72	290	0.931	735	615
0.366[4]	17.39	5.28[6]	267	1.091	794	654

Notes: [1]Upper yield, [2]Lower yield and 0.2% offset yield. [3]Ultimate. [4]Fracture. [5]Calculated from $(1 + \varepsilon)$ where d is not measured. [6]Measured from the broken specimen.

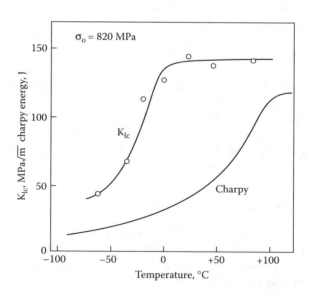

FIGURE 7.21 Temperature dependency of fracture toughness for component.

can be up to 400 MPa, and the operating temperature range is from −50°C to + 50°C. Assuming a half-circular crack in an effectively infinite mass of material, determine the crack size that you will need to be able to detect at each of the extreme values of operating temperature, if a factor of safety of 2 is required against brittle fracture. Comment on the results and describe NDE methods that you would recommend to detect cracks of these sizes.

7.3 The following data were obtained from Charpy impact tests at different temperatures on samples of a particular type of steel:

Temperature (°C)	Charpy Test Impact Energy (Joules)
50	76
40	76
30	71
20	58
10	38
0	23
−10	14
−20	9
−30	5
−40	1.5

1. Determine the ductile-to-brittle transition temperature.
2. Estimate the fracture toughness from the impact energy at 20°C and at 0°C.
3. If a large plate of the material with a yield strength of 1500 MPa is subjected to a load of 1000 MPa, calculate the critical crack size at a temperature of 20°C and at 0°C.

REFERENCES

1. Davis, J.R. (Ed.), *Metals Handbook — Desk Edition*, ASM, OH, 1998. pp. 119–121, pp. 1317–1321.
2. Cech, J., Measuring the mechanical properties of cast irons by NDT methods, *Nondestr Test Eval Int* 23, 93, 1990.
3. Schulz, A.W., Bonded joints and non-destructive testing: aiding the designer and user of fibre composites, *NDT Int* 4, 313, 1971.
4. Meyers, M.A. and Chawla, K.K., *Mechanical Behavior of Materials*, Prentice Hall, Upper Saddle River, NJ, 1999.
5. Cawley, P., The sensitivity of the mechanical impedance method of nondestructive testing, *Nondestr Test Eval Int* 20, 290, 1987.
6. Betz, C., *Principles of Penetrants*, Magnaflux Corp., Chicago, IL, 1969.
7. de Sterke, A., A practical introduction to penetrants, *NDT Int*, 1, 306, 1968.
8. Iddings, F.A., Visual inspection, *Mater Eval* 56, 816, July 1998.
9. Lomerson, E.O., Liquid penetrants, *Mater Eval* 36, 22, 1978.

10. Iddings, F.A., The basics of liquid penetrant testing, *Mater Eval* 44, 1364, November 1986.
11. *Standard Test Method for Liquid Penetrant Examination*, E165-95, Annual Book of ASTM Standards, Vol. 03.03 Nondestructive Testing, ASTM, West Conshohcken, Pennsylvania, 2001, pp. 60–79.
12. Sherwin, R.G. and DuBosc, P., Trends affecting penetrant specifications and standards: beyond 2000, *Mater Eval* 55, 864, August 1997.
13. Bailey, W.H., Millennium liquid penetrant testing, *Mater Eval* 58, 940, August 2000.
14. Thompson, R.B. and Brasche, L.J.H., Nondestructive evaluation of aircraft turbine engines, (Journal/volume/page/date coordinates?).
15. Becker, W.T. and Shipley, R.J. (Eds.), Practices in failure analysis, *ASM Handbook: Failure Analysis and Prevention*, Vol. 11, ASM, Materials Park, OH, 2002, p. 393.
16. Fomitchov, P., Wang, L.S., and Krishnaswamy, S., Advance image-processing techniques for automatic nondestructive evaluation of adhesively-bonded structures using speckle interferometry, *J NDE* 16, 215, 1997.

FURTHER READING

Felback, D.K. and Atkins, A.G., *Strength and Fracture of Engineering Solids*, Prentice Hall, Upper Saddle River, NJ, 1996.
http://www.ndt-ed.org/EducationResources/CommunityCollege/PenetrantTest/cc_pt_index.htm.
Lovejoy, D., *Penetrants Handbook*, Castrol Ltd., Swindon, U.K., 1990.
Larson, B.F., Study of the Factors Affecting the Sensitivity of Liquid Penetrant Inspections: Review of Literature Published from 1970 to 1998, FAA Technical Report Number DOT/FAA/AR-01/95, Office of Aviation Research, Washington, D.C., January 2002
Lovejoy, D., *Capabilities and limitations of NDT. Part 2: Penetrant methods*, British Institute for Nondestructive Testing, Northampton, 1989.
Tracy, N.A., *Handbook of Nondestructive Testing: Vol. 2. Liquid Penetrant Testing*, ASNT, Columbus, 1999.

8 Ultrasonic Testing Methods

This chapter looks at the various ways that ultrasound can be utilized for non-destructive evaluation of materials, including bulk and surface waves. It also includes discussion of the different ways that ultrasound can be generated in a material and consideration of the factors that affect signal amplitude — both those resulting from defects and those resulting from the normal attenuation with distance of signals passing through materials. The effects of inhomogeneities and interfaces on ultrasonic waves are considered, and the resulting reflection and refraction of wave front.

8.1 GENERATION OF ULTRASOUND IN MATERIALS

Ultrasound can be generated over a wide range of frequencies — from 20 kHz up to the MHz regime. At the low-frequency end of this range, the measurement techniques for determination of ultrasonic velocity depend on resonance using a continuous wave method, meaning that the excitation of the ultrasound in the specimen is not terminated. At the upper end of the frequency range, the techniques involve what is known as "pulse-echo" methods, in which a short duration burst of ultrasound is launched into the sample and the wave front passes down the sample. After reflection from an internal or external surface, it returns to the transducer as an echo. Normally, the sound is generated by ultrasonic transducers [1], but in the case of acoustic emission [2], the sound is generated by the growth of crack in the material.

8.1.1 TRANSDUCERS

To get ultrasound into a specimen, there has to be a device that converts electrical energy into acoustic energy. In some ways, this is analogous to a loudspeaker in the acoustic frequency range. In ultrasonics, it is called a transducer and is comprised most often of a piece of piezoelectric material, as shown in Figure 8.1, or less often, a piece of magnetostrictive material. In Figure 8.1 (a) the two electrodes are on opposite faces on the disk. In Figure 8.1(b), the electrodes are arranged coaxially with the first electrode at the center of the upper face, while the second electrode is arranged circumferentially on the upper surface and extends along the edges to the lower surface. The electric field lines in the transducer are also shown.

The transducer is normally enclosed in a housing, as shown in Figure 8.2. The piezoelectric material with its two electrodes is sandwiched between two

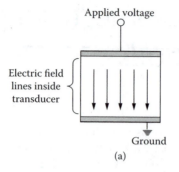

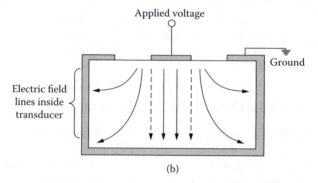

FIGURE 8.1 Two configurations for making electrical connections to a piezoelectric transducer. The piezoelectric material is in the form of a disk viewed "edge on." This is subjected to an alternating voltage of the required frequency supplied to the electrodes on its surfaces.

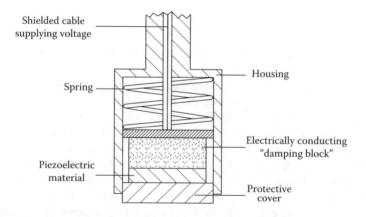

FIGURE 8.2 Schematic showing the housing of a piezoelectric transducer for use in pulsed ultrasonic mode.

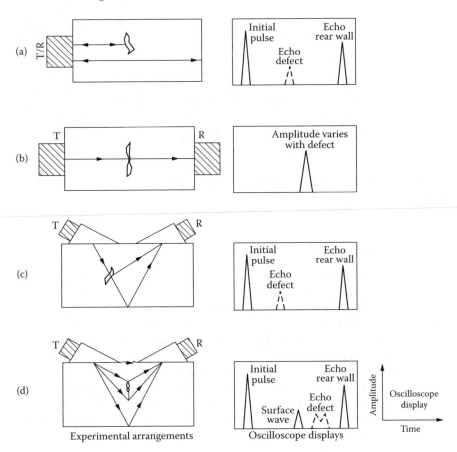

FIGURE 8.3 Various configurations of transducers commonly used in ultrasonic inspection: (a) single-ended transducer (pulse-echo), (b) double-ended transducers (pitch-catch) with through transmission, and two examples of doubled-ended transducers (pitch-catch) with angle beams (c) and (d).

blocks of non-piezoelectric material: the protective cover, and a conducting "damping block" that alters the resonance frequency of the transducer. The electrodes are connected via a coaxial cable to an AC voltage supply that is usually operated at the resonant frequency of the transducer. The transducer is encapsulated in a rigid protective housing and is spring-loaded to improve acoustic coupling on the surface of the test material.

The main configurations for transducers used to launch and detect waves for ultrasonic inspection are shown in Figure 8.3.

Pulse-echo ultrasonic inspection uses high-frequency (MHz) ultrasound to determine distance to a feature through measurement of time of flight. Most often, this uses a single transducer and relies on being able to detect the wave reflected from the opposite side of the sample (the back wall reflection), or the reflection from a flaw. For normal incidence, the depth of the flaw relative to the back wall is proportional to the time of flight of the flaw signal relative to the time of flight of the reflection from the back wall.

A two-transducer method can also be used where access to the far side of the sample is possible. This is less widely used than the pulse-echo arrangement. A third method, known as "pitch-catch", uses two transducers (a transmitter and a receiver) with angle beams of ultrasound, which is described in the following subsection. This arrangement can have advantages for getting the sound beam into "difficult" regions of a material in which the use of normal incidence beams is precluded.

Examples of the different physical effects that can form the basis of an ultrasonic transducer are shown in Table 8.1.

Examples of different transducer materials are shown in Table 8.2.

8.1.2 BEAM DIVERGENCE

When a transducer that is not a point source emits a wave, the wave front close to the transducer moves as a plane wave front, but at greater distances the beam diverges. In a simplified description of this effect, we talk about a "near zone" (also known as the Fresnel zone) where the beam is approximately parallel, and a "far zone" (also known as the Fraunhofer zone) where the beam diverges. These are shown in Figure 8.4.

The extent of the near-field zone, l, is dependent only on the diameter, D, of the planar transducer and the wavelength, λ, of the ultrasonic signal, assuming that the diameter of the transducer is much larger than the wavelength of the ultrasound (otherwise, the transducer is more like a point source than an extended planar source),

$$\ell = \frac{D^2}{4\lambda}. \tag{8.1}$$

For a given size of transducer, the longer the wavelength, the shorter the near-field zone. In the far-field zone, the beam diverges as if it had come from the center of the transducer; that is, it behaves in the far field as though it had come from a point source, with an angle of divergence, α, given by

$$\sin \alpha = \frac{1.22\lambda}{D}. \tag{8.2}$$

The width of the beam at a distance x is $2x \tan (\alpha/2)$.

8.1.3 DISTANCE–AMPLITUDE CORRECTION CURVE

Both attenuation and beam divergence cause a reduction in the intensity of ultrasound with distance. Therefore, a method is needed to enable us to compare

TABLE 8.1
Ultrasonic Transducers

	Materials	Frequencies
Piezoelectric and Ferroelectric Effect		
When crystals with asymmetric structures are subject to pressure, electrical charges develop on opposing crystal faces, and the crystal changes size. The inverse also occurs where crystals change size when subjected to an electric potential.		0.1–25 MHz
Magnetostriction Effect		
Change in dimension of ferromagnetic materials on magnetization. (Very limited in NDT application.)	Ferromagnetic materials, Ni	<200 kHz (upper limit determined by heating of specimen)
Electromagnetic Acoustic EMA, EMAT		
Induced eddy currents at radio frequencies in an electrical conductor in a magnetic field cause the surface to vibrate.	Electrical conductors	~2 MHz
Laser Generation		
Localized heating by laser pulses causes material to expand and contract very rapidly.	All solids and liquids	—
Others		
Several other physical effects are used in transducers to produce ultrasonic waves.	—	—

the significance of the strength of signals that have traveled different distances. For example, the signal that is a reflection from a flaw that is nearby is likely to be larger than the signal from a flaw that is farther away. This may be true even if the flaw that is farther away is actually physically larger. Therefore, a means to compare the signals is needed that takes into account the differences in distance that the signals have traveled. In this way, the signals can be "normalized" and compared on the basis of their significance.

The correction that is used to normalize the reflected signal from flaws at different distances is known as the distance–amplitude correction curve (DACC) [3]. When the ultrasonic signal is displayed on an oscilloscope, the time delay of the echo is a

TABLE 8.2
Materials for Ultrasonic Transducers

| Material | Chemical Formula | Material Type | Piezoelectric Constants | | Specific Acoustic Impedance ($z = \rho V$), 10^6 kg/m².s |
			α mV$^{-1} \times 10^{-12}$	β mV^{-1} Pa$^{-1} \times 10^{-3}$	
α-quartz	SiO₂	Piezoelectric single crystal	2.3	58	15.2
Barium titanate	BaTiO₃	Ferroelectric polycrystalline ceramic	149	14	29–31
Lead zirconate titanate (a)	PbZrO₃–PbTiO₃ solid solution PZT (PZT5)	Ferroelectric polycrystalline ceramic	374	15	28–30
Lead metaniobate(a)	PbNb₂O₆PMN	Ferroelectric polycrystalline ceramic	85	43	16–21
Lithium sulfate (hydrated)	Li₂SO₄LSH	Piezoelectric single crystal	16	175	11.2
Polyvinylidene fluoride	[CH₂-CF₂]ₙ PVDF	Ferroelectric plastic	—	—	4.1
Rochelle salt	KNaC₄H₄O₆·4H₂O	Piezoelectric single crystal	—	—	—
Backing materials (resins)	—	—	—	—	3–5

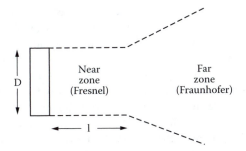

FIGURE 8.4 Near and far zones showing where a wave front emitted by an extended plane source is either a planar wave front (near-field zone), or a divergent beam (far-field zone).

representation of distance, so the DACC can be applied directly to the signals on the oscilloscope screen to make a comparison of significance. An example is shown in Figure 8.5.

In the idealized case of an exponential decay of signal, the amplitude can be represented by the equation

$$A(x) = A(0)\exp(-\mu x), \tag{8.3}$$

where $A(x)$ is the amplitude of the signal at a distance x, $A(0)$ is the amplitude at $x = 0$, and μ is the exponential decay coefficient. Because $x = vt$, this can be written

$$A(t) = A(0)\exp(-\alpha t), \tag{8.4}$$

where $\alpha = \mu v$. In general, the decay will not be truly exponential, but in some cases it will be, for example, if the beam is not diverging. Furthermore, it will not be possible in most cases to know α in advance; therefore it is necessary to measure

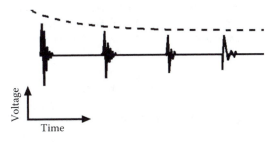

FIGURE 8.5 Example of a typical decay of ultrasonic echo amplitude with distance/time that appears on an oscilloscope screen.

the relative amplitudes of echoes from identical flaws at different known depths inside the material. This usually involves the measurement of signals from calibration blocks with a number of flat-bottomed holes at different depths below the surface [4]. From these results it becomes possible to determine the value of the decay coefficient α, and thereby to draw the DACC as a function of distance.

The attenuation coefficient μ can vary dramatically from one type of material to another. For example, μ is generally much larger in polymers than in metals. This means that ultrasonic echo signals decrease rather more quickly with both time and distance in polymers than in metals. Therefore, polymers can pose more problems than metals for detection of distant defects, and more careful testing is needed to detect these defects in polymers [5,6].

WORKED EXAMPLE

Four echoes are received from an original pulse of ultrasound that was launched into a sample from a single-ended transducer:

TABLE 8.3

	t (μsec)	$A(t)$ (mV)
A	10	0.8
B	15	0.7
C	30	0.6
D	50	0.5

If the amplitude of the original signal at $t = 0$ is 1 mV, determine which of these echoes is the most significant in terms of expected flaw size, then put the others in order of significance.

Solution

Assuming that the decay of the amplitude of the signals should be exponential, it is possible to determine the effective decay coefficient for each echo. The echo that has decayed least relative to the exponential expectation is the one with the lowest value of α. This then is the most significant and, assuming that the normalized amplitude of the signal is proportional to the size of flaw, it represents the largest flaw,

$$\alpha = \frac{1}{t} \log e \left(\frac{A(0)}{A(t)} \right). \tag{8.5}$$

Let us take the first echo and determine the nominal decay coefficient,

$$\text{A 10 μsec: } \alpha = \frac{1}{t} \log_e \left(\frac{A(0)}{A(t)} \right) = \frac{1}{10 \times 10^{-6}} \log_e \left(\frac{1.0}{0.8} \right) = 0.022 \ (\mu s)^{-1} \tag{8.6}$$

and determine the attenuation coefficients for the other echoes:

$$B \ 15 \ \mu sec: \ \alpha = 0.024$$
$$C \ 30 \ \mu sec: \ \alpha = 0.017 \qquad (8.7)$$
$$D \ 50 \ \mu sec: \ \alpha = 0.014$$

Therefore, on the basis of a correction using an exponential DACC, the echo signal at 50 μsec is the most significant, followed by the signal at 30 μsec, then the signal at 10 μsec, and, at last, the signal at 15 μsec is the least significant on the basis of this analysis.

An alternative solution to this problem is to assume a nominal decay coefficient based on the first echo and then normalize the detected echo amplitudes using the exponential equation. The nominal decay coefficient is then utilized to obtain the amplitudes for each signal at $t = 0$,

$$A(0) = A(t)\exp(\alpha t). \qquad (8.8)$$

This, then, allows a direct comparison of the strengths of the signals with the differences in distance/time having been eliminated.

$$A \ 10 \ \mu sec: \ A(0) = 0.8\exp(0.022 \times 10) = 1.00 \ mV \qquad (8.9)$$

$$B \ 15 \ \mu sec: \ A(0) = 0.7\exp(0.022 \times 15) = 0.97 \ mV \qquad (8.10)$$

$$C \ 30 \ \mu sec: \ A(0) = 0.6\exp(0.022 \times 30) = 1.16 \ mV \qquad (8.11)$$

$$D \ 50 \ \mu sec: \ A(0) = 0.5\exp(0.022 \times 50) = 1.50 \ mV \qquad (8.12)$$

These normalized amplitudes at $t = 0$ can then be compared directly, and they show that the 50 μsec signal is largest, followed by the 30 μsec, then by the 10 μsec, and finally the 15 μsec signal.

8.1.4 DISPLAY AND INTERPRETATION OF ULTRASONIC DATA

Ultrasonic data from the pulse-echo method can be displayed in a variety of forms, depending on what information is required from the data. These are systematically codified as A-scan, B-scan, and C-scan, depending on the form of representation.

A-scan amplitude vs. time: The A-scan is a plot of ultrasonic signal amplitude in volts against time, as shown in Figure 8.6. By convention, the amplitude is shown on the y-axis of an oscilloscope and the time is shown along the x-axis. A simple arrangement for making A-scans in a test specimen can be set up using a single-ended transducer method.

B-scan amplitude vs. depth: The amplitude does not need to always be plotted against time, as it is on an oscilloscope. Often, it is much more

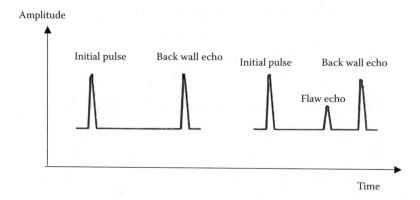

FIGURE 8.6 Schematic of an A-scan showing the variation of signal amplitude with time.

useful to know where high-amplitude reflections are coming from and the timing of these reflections, which correspond to depth. Measurement of the transit time and knowledge of ultrasonic velocity in the test specimen, allow the distances to any feature, such as those indicated by any unexpected echoes, to be calculated. The result is a map showing at what depth defects are located within a sample (Figure 8.7).

C-scan amplitude vs. position: Another way to look at the data is to find the locations of regions of high-amplitude reflections underneath the surface of a material. The result is then a map of the locations of defects within the test specimen (Figure 8.8).

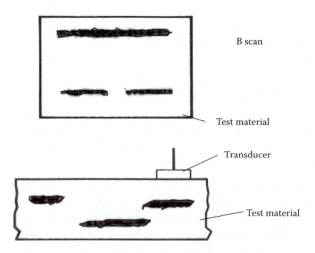

FIGURE 8.7 Schematic of a B-scan showing the depth of defects within a test specimen.

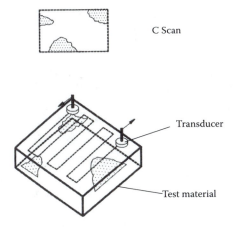

FIGURE 8.8 Schematic of a C-scan showing the locations of defects within a test specimen.

8.1.4.1 Interpretation of Ultrasonic Pulse-Echo Signals

It is worth remembering that not all ultrasonic signals correspond to defects or flaws [7]. As an example, in Figure 8.9 it can be seen that after the initial pulse there are two echoes, the first of which corresponds to a defect within the

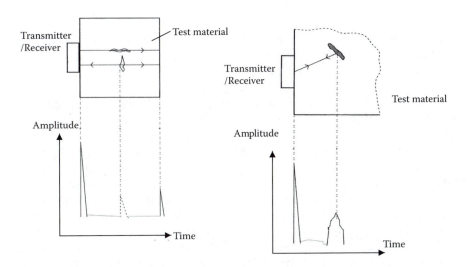

FIGURE 8.9 Oscilloscope display of an ultrasonic pulse train showing the initial pulse, the reflected echo from a defect, and the reflection from the back surface.

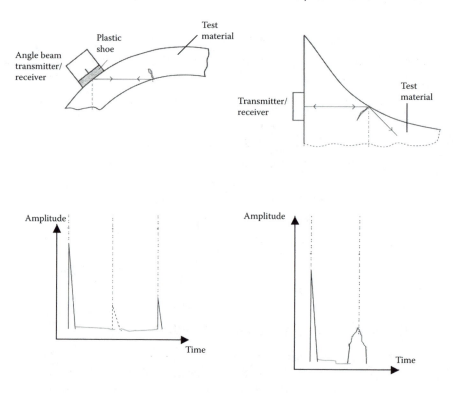

FIGURE 8.10 Oscilloscope display of an ultrasonic pulse train showing the initial pulse and the echo reflected from a defect.

material that is located about halfway between the front and back surfaces of the sample. The second echo comes from the back surface of the material and is known conventionally as the *back wall echo*. This does not correspond to a defect or flaw.

There are also situations where there are no back wall reflections, as shown in Figure 8.10. In this case, the only echoes that are seen are the reflections from defects or flaws.

Interpretation of these ultrasonic pulse-echo signals can be difficult. As mentioned above, the farther the ultrasound has traveled in the material, the greater the attenuation and, hence, the smaller the signal. This can be accounted for, at least approximately, by the DACC. Further interpretation can be complicated by the fact that echoes may be received as a result of multiple reflections, or that there may not be a back wall reflection to provide a reference for determining whether certain echoes are coming from reflections off surfaces that are closer than the back wall. In fact, when using a single-ended inspection without a back wall reflection, it can be quite difficult to interpret signals.

Thickness measurements depend on "time-of-flight" analysis, assuming that the ultrasonic velocities in the various components are known [8].

8.2 INHOMOGENEOUS AND LAYERED MATERIALS

So far everything that has been discussed in connection with ultrasonic inspection assumes that the material is homogeneous, and that any deviations from homogeneity, such as may be indicated by internal reflections of ultrasonic signals, are indicative of problems. This is not necessarily correct, because many modern materials are meant to be inhomogeneous, which will result in internal reflections of ultrasonic waves. The simplest case to analyze is that of layered materials.

Ultrasonics has been used for many years as one of the principal methods of nondestructive testing and evaluation of defects in materials. This is usually achieved by measuring the time taken for backscattered waves to return to a transducer to determine the location, and then measuring the change in amplitude of the backscattered signal [9] to estimate the size of defect. Laser ultrasound can be used to detect and characterize planar defects in materials [10] and has the advantage of being a noncontact method of evaluation.

Ultrasonics can be used for simultaneous velocity, thickness, and profile imaging [11]. In many cases the thickness and velocity are unknowns, so that variations in time of flight do not by themselves give a good indication of the state of the test specimen. Ultrasonic measurements can also be used for detection of stress in both small and large structures, as in the recent example of detection of stress in bridges [12], in which the strain, $\Delta\varepsilon$, is proportional to the fractional change in time of flight of ultrasound.

WORKED EXAMPLE

A reinforced structure 10 cm thick consists of lead on nickel, but the thicknesses of the layers are unknown. The time taken for a sound wave to travel through both layers of material is 40 msec. If the velocity in lead is 2.16 km.sec^{-1} and the velocity in nickel is 5.63 km·sec^{-1}, find the thickness of each of the two layers.

Solution

If the thicknesses of the two layers are dL and dN, then

$$dL + dN = 0.1, \tag{8.13}$$

and from the time of flight

$$t = \frac{dL}{2.16 \times 10^3} + \frac{dN}{5.63 \times 10^3} = 40 \times 10^{-6}. \tag{8.14}$$

Solving these two equations simultaneously gives

$$dL = 0.078 \text{ m} \tag{8.15}$$

$$dN = 0.021 \text{ m} \tag{8.16}$$

We now look at some of the complications that can arise if a material is layered.

8.2.1 TRANSMISSION AND REFLECTION AT INTERFACES

Impedance, in general terms, is the resistance of a material to the flow of any form of energy. This could be electrical energy, heat energy, or acoustic/ultrasonic energy. In the familiar case of electrical energy, impedance is calculated by dividing the voltage difference ("electrical pressure") across the ends of a material by the current flowing though the material (the product of the velocity of the charge carriers and the charge that they carry). Acoustic impedance, by analogy, is the ratio of the particle velocity to the pressure applied across the sample. In general, the following equation can be used to calculate the acoustic impedance Z,

$$Z = \rho V, \tag{8.17}$$

where V is the acoustic velocity, and ρ is the density. Ultrasonic waves have different velocities and, hence, the acoustic impedance is different in different materials, as shown in Table 8.4.

If we consider a multilayered structure, as shown in Figure 8.11, then, because of the different acoustic impedances of the various layers, reflection of ultrasound will occur at the interfaces between these layers.

8.2.2 AMPLITUDE OF REFLECTED WAVE

Consider the interface between two materials with acoustic impedances Z_1 and Z_2. For normal (perpendicular) reflection, the ratio of reflected to incident amplitude is

$$\frac{A_R}{A_I} = \frac{Z_1 - Z_2}{Z_1 + Z_2}, \tag{8.18}$$

where $Z_1 = \rho_1 V_1$ and $Z_2 = \rho_2 V_2$.

8.2.3 ENERGY TRANSFER AND CONSERVATION

The energy transfer across an interface depends on the square of the amplitudes, which is the intensity I of the wave. The ratio of reflected energy is therefore the ratio of intensities

$$\frac{I_R}{I_I} = \left(\frac{Z_1 - Z_2}{Z_1 + Z_2} \right)^2. \tag{8.19}$$

The ratio of transmitted energy is then determined on the basis of conservation of energy, assuming that no energy is actually dissipated in the process.

TABLE 8.4
Longitudinal Acoustic Velocity and Impedance of Various Metals

Material	Density (ρ), g/cm³	Longitudinal Wave Velocity, km/s	Longitudinal Wave Impendance (Z), 10⁶ Rayl	Shear Wave Velocity, km/s	Shear Wave Impendance (Z), 10⁶ Rayl	Surface (Rayleigh) Wave Velocity, km/s	Surface (Rayleigh) Wave Impendance (Z), 10⁶ Rayl
Aluminium	2.7	6.35	17.1	3.1	8.4	2.8	7.6
Beryllium	1.85	12.5	23.1	8.71	16.0	7.87	14.6
Brass	8.1	4.2–4.7	34–38	2.12	17.2	1.95	15.8
Bronze	8.86	3.53	31.2	2.23	19.8	2.01	17.8
Bismuth	9.8	2.2	21.5	1.1	10.8	—	—
Cadmium	8.6	2.8	24	1.5	12.9	—	—
Copper	8.93	4.66	42	2.26	20.1	1.93	17.2
Gold	19.3	3.2	62	1.2	23	—	—
Indium	7.31	2.5	18.3	—	—	—	—
Lead	11.4	1.96	22	0.7	8	0.63	7.2
Magnesium	1.74	5.79	10	3.1	5.4	2.87	5.0
Molybdenum	10.1	6.29	63.5	3.35	33.8	3.11	31.4
Nickel	8.84	5.7	50	2.96	26.1	2.64	23
Niobium	8.57	4.92	42.2	2.1	18.0	—	—
Platinum	21.4	4.15	89	1.73	37	—	—
Silver	10.5	3.44	36	1.59	16.7	—	—
Mild steel	7.83	5.95	46.6	3.2	25.0	2.79	21.8
Stainless steel 347	7.89	5.79	45.7	3.1	24.5	—	—
Cast iron	7.7	3.5–5.9	27–45	2.4	18.5	—	—
Tantalum	16.6	4.1	68	2.9	48	—	—
Tin	7.3	3.32	24.2	1.67	12.2	—	—
Titanium	4.54	6.1	27.7	3.12	14.2	2.8	12.7
Tungsten	19.25	5.18	100	2.87	55.2	2.65	51
Vanadium	6.03	6.0	36	2.78	17	—	—
Zinc	7.1	4.2	30	2.41	17	—	—
Zicronium	6.48	4.65	30	2.25	15	—	—
Chromium	7.0	6.65	46.6	4.03	—	—	—

By conservation of energy,

$$I_R + I_T = I_I, \tag{8.20}$$

and therefore,

$$\frac{I_T}{I_I} = 1 - \frac{I_R}{I_I} = 1 - \left(\frac{Z_1 - Z_2}{Z_1 + Z_2}\right)^2 = \frac{4Z_1 Z_2}{\left(Z_1 + Z_2\right)^2}. \tag{8.21}$$

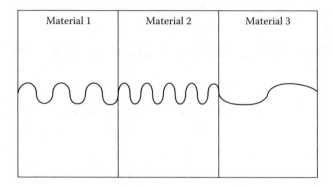

FIGURE 8.11 Multilayered structure in which each layer has a different acoustic wavelength and therefore a different acoustic impedance. (Note that these layers could also have the same wavelength but different densities and also still have different acoustic impedances.)

An important check here is to verify what happens when $Z_1 = Z_2$. In this case, all incident energy is transmitted, because in effect there is no acoustic interface.

8.2.4 DEPENDENCE OF REFLECTED ENERGY ON IMPEDANCES

The intensity of the reflected beam depends on the ratio of impedances of the two regions on either side of the interface. When the ratio of impedances is 1, then there is no reflected beam, so that $I_R/I_I = 0$ under this condition. As the ratio of impedances increases, the reflection ratio I_R/I_I approaches 1 asymptotically. This is shown in Figure 8.12.

8.2.5 AMPLITUDE OF TRANSMITTED WAVE

It is usually a reasonable assumption that there is no absorption of energy at the interface, and so the concept of conservation of energy applies to the incident, transmitted and reflected waves as shown in equation (8.20). Remember, however, that this does not mean that amplitude is conserved:

$$A_I \neq A_R + A_T. \tag{8.22}$$

In fact, the transmitted amplitude can be calculated from the principle of conservation of energy at the interface using the previous equation,

$$\frac{A_T}{A_I} = \sqrt{\frac{I_T}{I_I}} = \frac{2\sqrt{Z_1 Z_2}}{Z_1 + Z_2}. \tag{8.23}$$

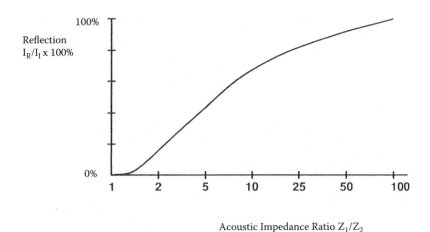

FIGURE 8.12 Ratio of reflected intensity to incident intensity as a function of the ratio of acoustic impedances.

8.3 ANGLE BEAMS AND GUIDED WAVES

So far we have talked only about normal or perpendicularly incident waves. These are waves that are launched at right angles to the free surface of a material under test. It is often necessary to use beams of ultrasound that are launched in directions other than the normal direction to reach regions of material that are at high risk or where it is considered that defects may be located. A good example of this difficulty occurs in welds, where the region at risk of cracking lies under the crown of the weld, and therefore is not accessible to normal beams. By suitable choice of the angle of incidence, it is possible to completely internally reflect some modes of the sound waves—for example, the longitudinal wave can be totally internally reflected if the correct angle of incidence is used.

WORKED EXAMPLE

A plate 0.02 m thick is inspected with a 70° angle beam. An echo indicating a discontinuity occurs at $t = 10$ μsec. If the velocity of sound is 8 km·sec^{-1}, what is the distance to the flaw along the surface of the plate?

Solution

Let D be the horizontal distance of the flaw from the transducer. The path length d of the ultrasonic wave in the material from time of launch until the detection of the echo is

$$2d = vt = 0.08 \text{ m} \tag{8.24}$$

$$d = 0.04 \text{ m} \tag{8.25}$$

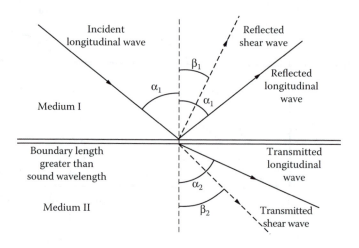

FIGURE 8.13 Wave mode conversion at an interface in which the incident longitudinal beam gives rise to shear reflected and refracted beams, in addition to the usual longitudinal reflected and refracted beams.

The horizontal distance along the surface is

$$D = D \cos (90 - 70) = D \sin (70)$$

$$= 0.0375 \text{ m}$$

(8.26)

The answer is the same no matter how many reflections of the ultrasound occur from the two surfaces.

8.3.1 MODE CONVERSION

Mode conversion means that some energy from an incident longitudinal wave is converted to a shear wave at the interface. This means that there can be two reflected beams and two refracted beams, one each of transverse and longitudinal vibrational modes. This is shown in Figure 8.13.

8.3.2 NONNORMAL REFLECTION

In nonnormal reflection — meaning, of course, any reflection that is not perpendicular to the surface — some energy is transferred and some is reflected. The amounts of acoustic energy reflected and transmitted depend on the angle of incidence and the ratio of acoustic impedances, and are different from those obtained at normal incidence. At sufficiently large angles of incidence, total reflection of the incident ultrasound can occur. This is shown in Figure 8.14.

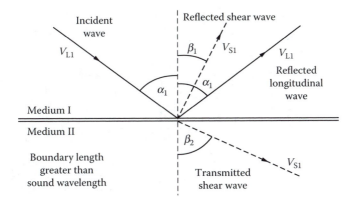

FIGURE 8.14 The angle of critical incidence, above which the refracted beam does not penetrate the second material.

8.3.3 REFRACTION

The angle of refraction depends on the incident angle and the velocities of sound in the two materials on either side of the interface. Assuming there is no mode conversion occurring, the angle of reflectance equals the angle of incidence, and the angle of refraction is related to the angle of incidence by Snell's law,

$$\frac{\sin \theta_1}{\sin \theta_2} = \frac{v_1}{v_2}, \tag{8.27}$$

where v_1 is the velocity in the first (incident) region, v_2 is the velocity in the second (refracted) region, θ_1 is the angle of incidence, and θ_2 is the angle of refraction. When the angle of refraction reaches 90°, the beam can no longer penetrate the second medium and this is known as *critical refraction*.

8.3.4 SURFACE ACOUSTIC WAVES

Apart from the usual bulk ultrasonic waves, such as longitudinal (compressive) waves and shear waves, there is another whole class of vibrations known collectively as "guided waves." These guided waves result from geometrical constraints of ultrasonic waves arising from the shape of the test specimen. Examples of different types of guided waves are: (1) plate waves, (2) cylindrical waves, and (3) rod waves, which are named after the different types of geometrical constraints. Two well-known examples of guided waves are Rayleigh waves and Lamb waves [13].

The Rayleigh wave is one particular example of a guided wave. Surface waves can be generated on a semi-infinite half space, using a transducer with an angle

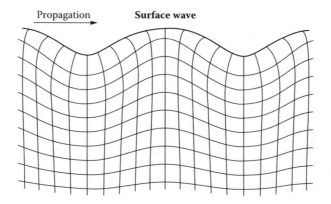

FIGURE 8.15 Rayleigh waves on the surface of a semiinfinite half space. (From Rindorf, H.J., Acoustic emission source location in theory and in practice, *Brüel and Kjær Technical Review*, No 2, 1981; Holford, K.M. and Lark, R.J., Acoustic emission testing of bridges, *in Inspection and Monitoring Techniques for Bridges and Civil Structures*, Fu, G., Ed., Woodhead Publishing Limited, Cambridge, 2005.)

beam as shown in Figure 8.3. These waves represent an idealized limiting case where the constraints imposed by boundary conditions are at a minimum. The solution modes of the wave equation in this case were first shown by Rayleigh [14] and included a surface traveling wave, now known as a Rayleigh wave, which is shown in Figure 8.15. In practical terms, the Rayleigh wave description applies only as an approximation to the vibrational modes of an arbitrarily thick plate.

In practice, all test specimens are of finite extent, and so a solution is needed in these more realistic cases. Another type of guided wave is the vibration of a plate of finite thickness. The solution in this case was first shown by Lamb [17] and the surface traveling wave that is generated is known as a Lamb wave, two examples of which are shown in Figure 8.16.

Guided ultrasonic waves are often used for nondestructive inspection of structures. One reason for this is that they travel with relatively low attenuation of the signal. The guided waves can be excited at one location, propagated over large distances, and detected at another location in pitch-catch mode. The detected signal contains information about the structure of the material between these locations through which the guided wave has passed [18]. In other cases, a single location for both excitation and detection can be used. This is the so-called pulse-echo mode.

Each type of guided wave in a structure has different vibrational modes. These modes represent different vibrational modes of the geometrical structure and are analogous, for example, to the numerous well-known vibrational modes of a beam, where the geometry of the beam is the dominant factor determining the possible vibrational modes. An advantage of the existence of these multiple modes is that each mode has a different sensitivity to particular types of defects in the structure. Therefore, by investigating the results for different modes,

Lamb waves
Symmetric, i.e. longitudinal in centreline

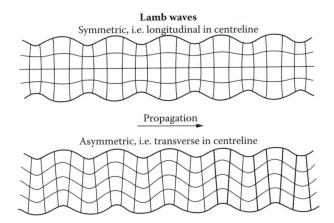

Propagation

Asymmetric, i.e. transverse in centreline

FIGURE 8.16 Two different forms of Lamb waves on the surface of a plate of finite thickness. (From Rindorf, H.J., Acoustic emission source location in theory and in practice, *Brüel and Kjær Technical Review*, No 2, 1981; Holford, K.M. and Lark, R.J., Acoustic emission testing of bridges, *in Inspection and Monitoring Techniques for Bridges and Civil Structures*, Fu, G., Ed., Woodhead Publishing Limited, Cambridge, 2005.)

characterization of defects is possible. Also, because the reflectance of guided waves from defects is different from reflectance of bulk waves, it is possible to detect defects that are much smaller than the wavelength of the guided wave.

In summary, Rayleigh and Lamb waves are particular examples of guided waves in structures. Because of the advantages of these types of waves for identification and characterization of defects as described above, they are receiving increasing attention for nondestructive evaluation and, in particular, Lamb waves are being used for nondestructive evaluation of plates [19].

EXERCISES: ULTRASONIC TESTING METHODS

8.1 Ultrasonic transducers project beams of ultrasound into materials, but these do not usually behave simply as plane waves. Explain what is meant by beam divergence, and give equations for the length of the near-field region (Fresnel zone) and the beam angle (angle of divergence) in terms of the diameter of a transducer and the wavelength of the ultrasonic wave.

A longitudinal wave transducer of diameter 25 mm and resonant frequency 5 MHz is used to inspect steel components. If the longitudinal wave velocity in steel is 5960 m/sec, what is the range of the near-field region and what is the beam width at 1 m from the transducer?

8.2 In testing a casting using ultrasound, an unexpected echo is obtained which is greater than full screen height on the oscilloscope. If the signal level is reduced by putting the attenuator setting at 10 dB, the flaw signal is just equal in height to a standard reference defect. What is the amplitude of the defect signal compared to the standard reference defect on a relative scale? If the size of the flaw suddenly increases, doubling the amplitude of the received signal, what change in attenuator setting is needed to bring the signal back to full height on the oscilloscope?

8.3 For a particular ultrasonic problem, it is required that only a shear wave is transmitted by mode conversion in a metal from a piezoelectric transducer generating longitudinal waves in a wedge-shaped polymer transducer "shoe." If the longitudinal velocity of the ultrasound in the polymer shoe is 2 km/sec, and the velocities of longitudinal and shear waves in the metal are 4.5 km/sec and 2.5 km/sec, calculate the range of angles of incidence that is needed.

REFERENCES

1. Silk, M.G., *Ultrasonic Transducers for Nondestructive Evaluation*, Adam Hilger, Bristol, 1984.
2. Richter, J.R., Acoustic emission testing: a composite manufacturers experience, *Mater Eval* 57, 492, May 1999.
3. Burkle, W.S. and Isom, V.H., Use of DAC Curves and transfer lines to maintain ultrasonic test references level sensitivity, *Mater Eval* 41, 624, May 1983.
4. Beck, K.H., Limitations to the use of reference blocks for periodic and preinspection calibration of ultrasonic inspection instruments and systems, *Mater Eval* 57, 323, 1999.
5. Sokolov, I.V., The split-method of ultrasonic nondestructive testing, *Nondestr Test Eval* 19, (page), 2003.
6. Ashton, J.N., Reid, S.R., Soden, P.D., Peng, C., and Heaton, M., Non-destructive evaluation of impact damaged composite pipes and pipe couplings with bond line defects, *Nondestr Test Eval* 19, 2003
7. Engblom, M., Considerations for automated ultrasonic inspection of in service pressure vessels in the petroleum industry, *Mater Eval* 47, 1332, December 1989.
8. Cartwright, D.L., Ultrasonic thickness measurements of weathering steel, *Mater Eval* 53, 452, 1995.
9. Mundry, E., Defect evaluation by ultrasonics: some results of work in progress at the Bundesanstalt fur Materialprufung, *Nondestr Test Eval Int* 5, 290, 1972.
10. Wooh, S.C. and Wang, J.Y., Nondestructive characterization of planar defects using laser-generated ultrasonic shear waves, *Research in NDE*, 13, 215, 2001.
11. Fei, D., Hsu, D.K., and Warchol, M., Simultaneous velocity, thickness and profile imaging by ultrasonic scan, *J NDE* 20, 95, 2001.
12. Fuchs, P.A. et al., Ultrasonic instrumentation for measuring applied stress on bridges, *J NDE* 17, 141, 1998.
13. Viktorov, I.A., *Rayleigh and Lamb Waves: Physical Theories and Applications*, Plenum Press, New York, 1967.
14. Lord Rayleigh (*née* J.W.Strutt), *Proc Lond Math Soc* 17, 4, 1885.
15. Rindorf, H.J., Acoustic emission source location in theory and in practice, *Brüel and Kjær Technical Review*, No 2, 1981.
16. Holford, K.M. and Lark, R.J., Acoustic emission testing of bridges, *in Inspection and Monitoring Techniques for Bridges and Civil Structures*, Fu, G., Ed., Woodhead Publishing Limited, Cambridge, 2005.

17. Lamb, H., On waves in an elastic plate, *Proc R Soc Lond Ser. A* 93, 114–128, 1917.
18. Balasubramaniam, K., *J Nondestr Test Eval* 1, xxx, 2002.
19. Alleyne, D.N. and Cawley, P., The interactions of Lamb waves with defects, *IEEE Trans Ultrasonics Ferroelectr Frequency Control* 39, 381, 1992.

FURTHER READING

Achenbach, J.D., *Evaluation of Materials and Structures by Quantitative Ultrasonics*, CISM course, Lecture No.330. Springer-Verlag, New York, 1994.

Auld, B.A., *Acoustic Fields and Waves in Solids*, Robert Krieger, Malabar, FL, 1990.

Birks, A.S. and Green, R.E., *Handbook of Nondestructive Testing: Ultrasonic Testing*, Vol. 7, ASNT, Columbus, 1991.

Blitz, J. and Simpson, G., *Ultrasonic Methods of Nondestructive Testing*, Chapman and Hall, London, 1995. 264 pp.

http://www.ndt-ed.org/EducationResources/CommunityCollege/Ultrasonics/cc_ut_index.htm.

Krautkramer, J. and Krautkramer, H., *Ultrasonic Testing of Materials*, 4th ed., Springer-Verlag, Berlin, 1990.

Rindorf, H.J., Acoustic emission source location in theory and in practice, *Brüel and Kjær Technical Review*, No 2, 1981; Holford, K.M. and Lark, R.J., Acoustic emission testing of bridges, *in Inspection and Monitoring Techniques for Bridges and Civil Structures*, Fu, G., Ed., Woodhead Publishing Limited, Cambridge, 2005.

Silk, M.G., *Capabilities and Limitations of NDT. Part 5: Ultrasonic Testing; Special Techniques*, British Institute for Nondestructive Testing, Northampton, 1989.

9 Electrical Testing Methods

This chapter looks at the use of electrical methods for evaluation of materials, particularly the use of eddy currents. Eddy current inspection depends on inducing time-dependent currents in materials, the strength of which depend on permeability and conductivity of a material. Thus, eddy current methods can be used to determine changes in microstructure of a material and any materials properties that are related to microstructure. Therefore, variations in heat treatment, mechanical hardness, impurities or other differences in chemical content, and corrosion damage can all be detected by eddy current inspection. Flaws in a material also interrupt the flow of eddy currents, and so eddy current inspection is also sensitive to these features. Macroscopic cracks in materials can be detected through their effects on eddy currents However, microscopic damage — such as the accumulation of dislocations under fatigue damage — that has effects similar to microstructural changes rather than macroscopic cracks, also affects the eddy current signals.

9.1 BASICS OF EDDY CURRENT TESTING

An eddy current inspection system consists of the following: a source of alternating magnetic field, a sensor to detect the flux density generated by the field, and analytical electronics to convert the resulting signals into a form that is easily interpretable. Often, the source and sensor are the same in these systems.

A time-varying magnetic flux density, B, will cause a circulating electric field according to Faraday's law [1],

$$\nabla \times E = -\frac{\partial B}{\partial t}.$$

(9.1)

This is usually achieved using a surface coil. In its simplest form, the eddy current sensor consists of a single coil, as shown in Figure 9.1a. The induced current in the test material absorbs or dissipates energy from the source coil, leading to differences in the flux density. This causes differences in the electrical impedance of the sensor coil, depending on the condition of the test material, particularly the permeability and conductivity of the material and any local variations in these properties caused by defects, cracks, or changes in microstructure [2].

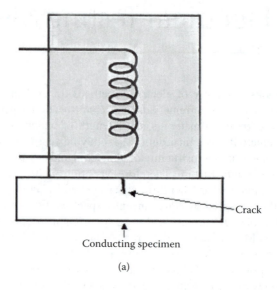

(a)

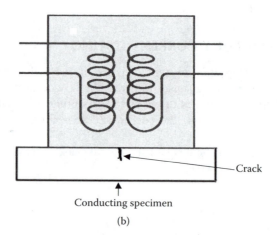

(b)

FIGURE 9.1 Eddy current inspection coils showing (a) a single coil sensor and (b) a differential coil sensor consisting of two identical coils connected in opposition.

9.1.1 Eddy Current Inspection

The eddy current response is simply a measure of the impedance of a coil subjected to a time-varying (usually sinusoidal) magnetic field when in the vicinity of a conducting material. The eddy currents generated in the material depend on the structure and properties of the material, so that measure of the impedance can be used to determine materials properties indirectly. Broadly, two types of eddy current tests are employed, which can be categorized as surface testing and tube testing [3].

The impedance of the coil, and particularly change in the impedance, is conveniently represented on an impedance plane which simultaneously shows the reactance and resistance of the sensor coil. This is usually represented on the screen of an oscilloscope by displaying the real and imaginary parts of the impedance. The vector representing these two quantities is known as the impedance phasor [1].

In many cases, the dielectric coefficient of a material is also structure sensitive, so the coupling between a flux coil and a test material in the vicinity of the coil depends on, among other factors, the structure of the material. Changes in the coupling alter the impedance of the flux coil. This means that we have an inspection method that can be used to investigate the structure of a material indirectly through the measurement of its permeability and permittivity.

If the primary objective is to detect cracks, it is often more effective to use a difference probe (sometimes called a differential probe), which has two nominally identical coils wrapped in opposition as shown in Figure 9.1b. When such a pair of coils is located over a homogeneous material, the two coils give the same signals, which then effectively cancel out because the coils are connected in opposition. If the pair of coils is located over an inhomogeneous material, then the signals will be different, so in effect there is only a net voltage signal from such a sensor when inhomogeneities (e.g., cracks) are present in the material.

9.1.2 IMPEDANCE PLANE RESPONSE

It is important to remember that in an eddy current sensor what is measured is the impedance of the coil [3]. Because it operates under AC conditions, the impedance of the coil is affected by the presence of conducting surfaces in the vicinity, to which the coil becomes coupled electromagnetically. The strength of this coupling, and hence the impedance of the coil, is affected by the proximity of the conducting surface. This proximity effect is often called the "lift-off" effect. Discontinuities in the surface also affect the impedance of the coil.

An analysis of the effects of lift-off on the impedance of air-cored eddy current coils has been presented [4] which shows a semiempirical relationship between impedance, frequency, and lift-off for coils of finite length. This provides a starting point for normalization of eddy current testing whereby the effects of lift-off can be eliminated from the measurements with these types of coils.

The lift-off effect is presented in Figure 9.2, which shows the impedance plane response of the coil when brought from a large distance (the so-called *air point*) toward the surfaces of different materials. These materials (copper, aluminum, brass, lead, titanium, stainless steel, and high-alloy steel) have different values of permeability and conductivity, and so their effects on the impedance of the coil are different. From such a curve, it is possible to calibrate the eddy current probe so that a test material can be identified uniquely from the end point (zero lift-off) on the impedance plane. However, if there is an unknown amount of lift-off from an unknown material, it becomes very difficult

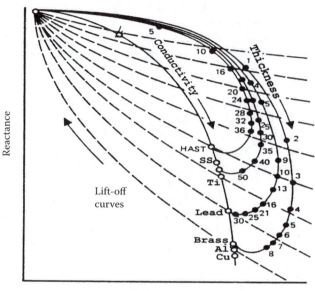

FIGURE 9.2 Lift-off curves for sheets of various metals stacked so that there is no further change in impedance with thickness when sensors are located directly in contact with the surface.

to identify the material. On the other hand, if the material is known, then it is possible to determine the lift-off distance from the impedance data.

The shape of the lift-off curve depends on the conductivity of the test material as shown in Figure 9.3.

The shape of the lift-off curve depends also on the frequency of excitation as shown in Figure 9.4.

As a result of all of these complications and possible factors that influence the actual response of the eddy current sensor, periodic calibration of the eddy current coil using known standards is recommended. Calibration before each set of measurements is desirable.

9.1.3 Depth Dependence of Intensity of Eddy Currents

An important question for eddy current inspection is how deeply the eddy currents penetrate into the material [5]. This will determine the depth of material through which information can be obtained by eddy current inspection. To answer this we need to know first what factors limit the penetration of a magnetic field into a material. The penetration depth depends on materials properties, such as conductivity and permeability, and also on external factors, such as the frequency of the magnetic field that is used to excite the eddy currents. Permeability, μ, is sensitive to structure of the material, including dislocations, residual stress, the presence of

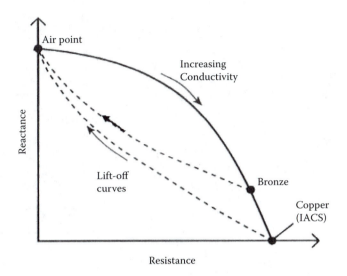

FIGURE 9.3 Conductivity curve and the lift-off curves of copper and bronze.

second phases (for example, in two-phase materials such as steels), and precipitates. Conductivity is sensitive to cracks, defects, voids, and any other factors that scatter the electrons as they move through the material.

The intensity of eddy currents changes as a function of depth, with a decrease, which in simplified cases, can be described by an exact equation. A good example is when a plane electromagnetic wave impinges on a flat surface, in which case

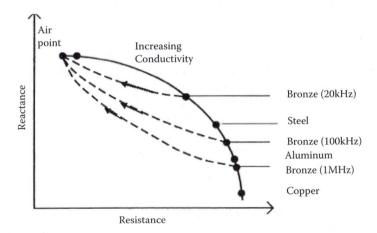

FIGURE 9.4 Frequency dependence of the impedance plane response for an eddy current coil close to conducting materials.

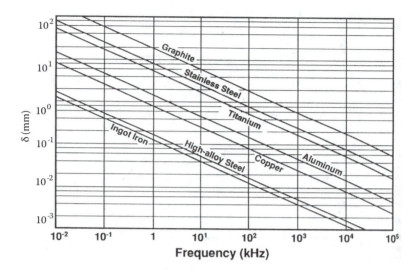

FIGURE 9.5 The variation of penetration depth with excitation frequency for various materials

an exact analytical mathematical expression can be obtained for the solution. At a fixed frequency, the field strength drops exponentially with the rate of decay determined by the penetration depth δ, so that

$$H = H_o \exp(-z/\delta). \tag{9.2}$$

Thus, at a depth $z = \delta$, the signal is at $1/e$ of its value at the surface.

The eddy currents penetrate into a material to a depth that may be considered to be typically three times the so-called skin depth. Therefore, if a test material has a thickness that is less than the skin depth, any variations in thickness can be detected as changes in the eddy current response. However, if the test material is thicker, then variations in thickness will have little effect on the eddy current signals. The method is therefore useful for thickness determinations of sheet metal and foils. In a related way, the eddy current method is sensitive to lift-off from the surface of a conducting material and can be used to determine the thicknesses of layers or coatings on the surfaces of conducting materials. This includes layers of paints or other insulating coatings.

9.1.4 DEPENDENCE OF PENETRATION DEPTH ON MATERIALS PROPERTIES

The penetration depth, δ, depends on the frequency of the excitation, ν, the permeability, μ, and conductivity, σ, of the material. For a plane electromagnetic wave impinging on a flat surface, the equation is

$$\delta = \sqrt{\left(\frac{1}{\pi v \sigma \mu}\right)}.$$ (9.3)

The variation of penetration depth, δ, for different materials and frequencies is shown in Figure 9.5.

9.2 EDDY CURRENT SENSORS

Eddy current sensors consist of a coil or coils that provide both generation and sensing functions. The coil is usually embedded at the end of a rigid cylinder in which the coil itself is protected from damage, usually by a polymeric filler, which keeps the coil rigid, thereby maintaining the same electrical parameters, such as cross-sectional area and length. A typical eddy current sensor configuration is shown in Figure 9.6.

9.2.1 VARIOUS SENSOR GEOMETRIES AND CONFIGURATIONS

The eddy currents can be generated in a test material through a variety of different geometrical configurations, as shown in Figure 9.7. Various setups for conducting eddy current inspection are also shown.

Although in principle these eddy current sensors are very simple, their performance in nondestructive evaluation of materials is dependent on geometry. In particular, their signal-to-noise ratios for detection of defects are sensitive to the details of the design. Chen and Miya [6] have developed an approach for design

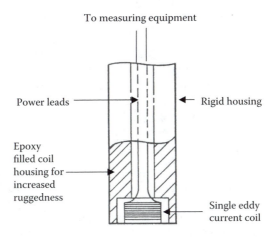

FIGURE 9.6 Eddy current inspection probe configuration. The coil may be a single coil or a pair of coils connected in opposition as shown in Figure 9.1.

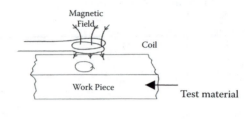

(a)

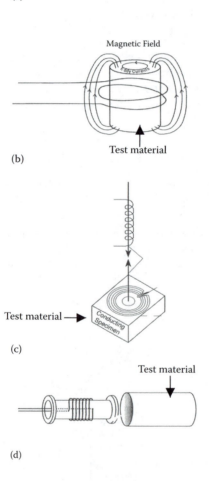

(b)

(c)

(d)

FIGURE 9.7 Various different configurations for eddy current inspection. These consist of (a) a single flat coil known as a pancake coil, suitable for inspecting flat surfaces, (b) an encircling coil, suitable for inspecting rod shaped specimens, (c) a pencil probe coil, suitable for inspecting small regions on a variety of surfaces with different curvatures, and (d) a bobbin coil suitable for inspection of tubes

of eddy current probes based on a simple, analytical "ring current" model, which can be used for approximately determining detectability of defects and signal-to-noise ratios for the probes.

9.2.2 Equivalent Circuits

The combination of the sensor coil and the test material can be described in terms of coupled equivalent circuits. The sensor's circuit can be described as a resistor, R_0, and an inductor, L_0, in series. Similarly, the test material can be described as a resistor, R_1, and an inductor, L_1. These two circuits are inductively coupled, and measured changes in the test material can be described in terms of changes in the two circuit parameters R_1 and L_1. The result is that the response of the combined sensor/test material combination can be described relatively simply in terms of the measured impedance Z, which has two components — a resistive component and a reactive component. The variations of these two measures with a variety of situations such as lift-off, cracks, and permeability and conductivity variations of the test material can then be calibrated.

9.2.3 Impedance Measurement

The detection electronics for eddy current sensors may vary, but the general concept is that it should consist of a Wheatstone impedance bridge. Balancing the bridge determines the impedance of the eddy current sensor coil. This impedance changes with the electromagnetic coupling between the sensor coil and the test material. This coupling is affected by materials properties and geometrical considerations, such as the physical separation of the coil (lift-off) from the test material.

It is normal for an eddy current sensor to be calibrated using a known reference specimen before going on to inspect the test specimen. However, Blitz et al. [7] have shown how the output of an eddy current sensor can be related directly to the depths of cracks in the test material.

9.2.4 Impedance Plane Representation

The electrical impedance of the eddy current coil varies under different test conditions, such as proximity to a test specimen, conductivity, and permeability of the test specimen, the presence of cracks or inhomogeneities, and frequency of excitation. The impedance has two components — the resistance, R, and the reactance, X_L ($= \omega L$). These are usually represented on an impedance plane. These impedance plane plots will usually show either the change in resistance, ΔR, plotted against the change in reactance, $\omega \Delta L$, or alternatively, the values of resistance and reactance normalized to the reactance of the coil, ωL_o, when it is far away from any test specimens — $R/\omega L_o$ and $\omega L/\omega L_o$. Because there are many variables that can affect the eddy current response, these plots are usually made with some of the variables held constant, such as measurement at a fixed distance (lift-off distance) and frequency for specimens with different combinations of conductivity/permeability, measurements on a particular test specimen at fixed frequency with different lift-off distances, or measurements on a particular test specimen at a fixed lift-off distance at different frequencies.

The generation of eddy currents in a test material affects both the resistance and reactance of the sensor coil when the sensor and test material are in sufficiently close proximity that they are electromagnetically coupled. The strength of coupling, and hence the strength of the eddy currents generated in the test material, are both affected by the physical separation of the sensor and the test material. Therefore, the response of the eddy current inspection is not solely dependent on the condition of the test material.

Because the sensor has an impedance when it is not in the vicinity of any test materials, this can be used as a baseline from which measurements of impedance can be compared. This baseline is known as the air point and is used as a convenient reference. The changes in impedance relative to this air point can be represented on an oscilloscope screen at much higher gain than the absolute values of impedance, and therefore, this gives a more sensitive measure of differences in the state of a test material. Some examples of changes in impedance as represented on an oscilloscope screen are shown in Figure 9.8.

The impedance vector can alternatively be expressed in terms of its magnitude, $|Z|$, and phase, ϕ, as shown in Figure 9.9,

$$|Z| = \sqrt{(R^2 + X^2)} \tag{9.4}$$

$$\phi = Arc\tan\left(\frac{X}{R}\right). \tag{9.5}$$

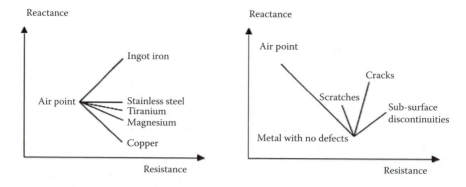

FIGURE 9.8 Vector representation of changes in complex impedance relative to the air point for various materials and for different types of defects. Z_0 is the air point and Z_1 is the impedance when over the metal with no defects present.

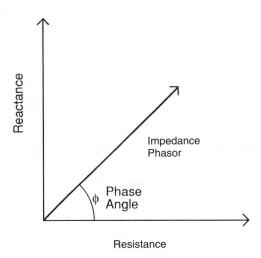

FIGURE 9.9 Magnitude and phase angle representation of complex impedance.

9.3   FACTORS AFFECTING EDDY CURRENT RESPONSE

The factors affecting the eddy current response can be separated into those that are related to the test material and those that are related to the sensor.

Factors related to the test material are the following:

- Conductivity
- Permeability
- Thickness
- Other geometrical factors (cracks etc.)

Factors related to the coil are as follows:

- Frequency
- Strength of electromagnetic coupling (mutual inductance)
- Coil geometry
- Proximity to test material (lift-off)

9.3.1   MATERIALS PROPERTIES AFFECTING EDDY CURRENTS

The following materials properties affect the eddy current signals. These factors can therefore be tested using eddy current inspection. It is usually difficult to interpret the eddy current signals when there are two or more parameters that vary.

Sample specific factors are listed as follows:

- Alloy composition
- Heat treatment
- Grain size
- Magnetic permeability
- Electrical resistivity
- Corrosion
- Temperature (because this affects both conductivity and permeability)

Geometrical factors comprise the following:

- Thickness
- Distance from other conductors
- Geometry of test sample
- Surface coating
- Lift-off
- Defects

The term *defects* is used here in its broadest sense, so that any unexpected or unwanted physical discontinuities in the material qualify as defects. This includes cracks, inclusions, dents, holes, and scratches.

9.3.2 Effects of Cracks on Eddy Currents

As already stated, the presence of cracks can alter the eddy current response quite markedly when they interrupt the path of eddy currents in the material. One technique that works very well in these cases is to use a differential sensor with two nominally identical coils wound in opposition, as shown in Figure 9.10. If the two coils are well matched, this has the advantage of only giving a signal when there is a defect or some other localized variation in properties present. In the case of such a differential sensor, it can be seen (from a symmetry argument) that the air point and a location on a perfectly homogenous sample both give the same zero output signal.

Another technique that has had success in the detection of defects and wall-thinning in tubular specimens is the remote field eddy current technique [8].

9.3.3 Geometrical Factors Affecting Eddy Currents

Because lift-off, conductivity, and thickness, all play a role in determining the eddy current response, these can be shown on the impedance plane representation, as depicted in Figure 9.11.

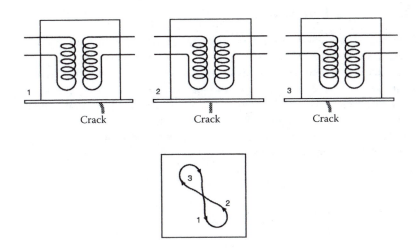

Crack   Crack   Crack

Corresponding oscilloscope trace

FIGURE 9.10 Schematic of a differential eddy current sensor showing how surface cracks produce a signal when the two coils are positioned asymmetrically with respect to the crack.

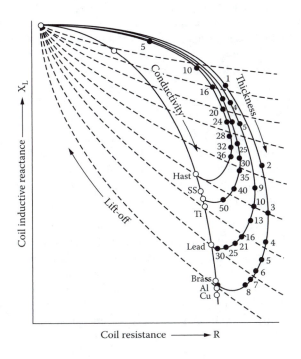

FIGURE 9.11 Effects of lift-off, conductivity, and thickness of test material on eddy current impedance plane response.

9.3.4 MUTUAL INDUCTANCE

When an eddy current inspection is performed there is no direct source of electric current in the test specimen. There are only induced currents. These arise because of the electromagnetic coupling between the sensor coil and the test material, which can be described in terms of the mutual inductance M_I. Returning to the earlier description of equivalent circuits, if R_1 and L_1 are the resistance and inductance of the equivalent circuit representation of the test material, then the change in resistance and inductance of the sensor circuit relative to the "air point" are given in terms of these circuit parameters, together with the mutual inductance and the frequency by

$$\Delta R = \frac{M_I^2 R_1 \omega^2}{R_1^2 + \omega^2 L_1^2} \tag{9.6}$$

$$\Delta L = \frac{M_I^2 L_1 \omega^2}{R_1^2 + \omega^2 L_1^2}. \tag{9.7}$$

Thus, for a particular test material and a given sensor operating at a set frequency, the factor determining the eddy current response is the mutual inductance, M_I, between the test circuit and the sensor. This depends on the permeability and conductivity of the test material, but also if the conduction path of the current in the material is interrupted, L_1 and R_1 can also be affected. This will result in a flaw indication in the eddy current signal.

EXERCISES: ELECTRICAL TESTING METHODS

9.1 A piece of nonferrous tubing is to be tested using eddy current methods by passing it through a coil carrying a current that is cycling at 70 kHz. The field produced by the coil extends over a distance of 4 mm. The design considerations require that a defect in the tube must be in the field of the coil for a minimum of 50 cycles. What is the maximum testing speed?

9.2 In air, an eddy current probe coil has a resistance of 20 ohms and a reactance of 20 ohms. Calculate the impedance of the coil in air. When placed on a sample of aluminum, the resistance increases by 10 ohms and the reactance decreases by 10 ohms. Calculate the impedance of the probe on aluminum.

When placed on stainless steel the changes in resistance and reactance, compared with the impedance in air, are +8 ohms and −5 ohms, respectively. If the conductivity of aluminum is 35×10^6 S/m and the conductivity of stainless steel is 2×10^6 S/m, describe how you would develop a calibration curve for determining conductivity of an unknown material. You may assume that the relative permeability of both materials is 1.

9.3 An eddy current inspection is performed on a test specimen at a frequency of 10 kHz. The equivalent circuit of the test material is an LR circuit with resistance $R = 20$ ohms and inductance 200 μH. Calculate the changes in resistance and reactance of the eddy current inspection probe as the coil is moved from a large distance (mutual inductance $M_I = 0$) until it touches the surface of the test material ($M_I = 100$ μH). With different test specimens the value of mutual inductance will change; plot the variation of reactance $\omega\Delta L$ with resistance ΔR for a series of different values of M_I. What do you notice?

REFERENCES

1. Libby, H.L., *Introduction to Electromagnetic Nondestructive Evaluation*, John Wiley & Sons, New York, 1971.
2. McMaster, R.C., The present and future of eddy current testing, *Mater Eval* 60, 27, 2002.
3. Birring, A.S. and Marshall, G.A., Eddy current testing in the petrochemical industry, *Mater Eval* 61, 1190, 2003.
4. Hajian, N.T., Blitz, J., and Hall, R.B., Impedance changes for air-cored probe-coils of finite lengths used for eddy-current testing, *Nondestr Test Eval Int* 16, 3, 1983.
5. Franklin, E.M., Eddy current inspection, *Mater Eval* 40, 1008, 1982.
6. Chen, Z. and Miya, K., A new approach for optimal design of eddy current testing probes, *J NDE*, 17, 105, 1998.
7. Blitz, J., Williams, D.J.A., and Tillson, J., Calibration of eddy current test equipment, *NDT Int* 14, 119, 1981.
8. Schmidt, T.R., History of the remote field eddy current inspection technique, *Mater Eval* 47, 17, 1989.

FURTHER READING

Auld, B.A. and Moulder, J.C., Review of advances in quantitative eddy current nondestructive evaluation, *J NDE*, 18, 3, 1999.

Blitz, J., *Electrical and Magnetic Methods of Nondestructive Testing*, Adam Hilger, Bristol, 1991.

http://www.ndt-ed.org/EducationResources/CommunityCollege/EddyCurrents/cc_ec_index.htm.

Udpa, S., *Handbook of Nondestructive Testing: Eddy Current Testing*, Vol. 5, ASNT, Columbus, 2004.

10 Magnetic Testing Methods

This chapter looks at magnetic methods. Magnetic methods are distinct from electromagnetic methods in that they depend on the relationship between structure and magnetic properties of materials. There are two broad classes of magnetic techniques: intrinsic (those that depend on the properties of the material) and extrinsic (those that depend on local variations in magnetic state of the material caused by defects such as cracks and other flaws). We look at how the magnetic field is generated in a material, how the bulk magnetization changes with magnetic field, and how understanding of this structure-sensitive relationship can be used for materials evaluation. We then look at how the local magnitude and direction of the magnetic field in a material is affected by defects, and how these local variations can be used to detect, locate, and classify defects.

10.1 MAGNETIZATION

When a material is subjected to a magnetic field, the response is magnetization. The measurable quantity is the magnetic flux density that arises from the combination of the magnetic field and the magnetization. The variation of flux density with field in strongly magnetic materials (ferromagnets and ferrimagnets) contains information on structure and state of stress of the materials and so can be used in materials evaluation [1].

The magnetization of a material can therefore be used for nondestructive evaluation. However, the existence of magnetized parts can be problematic for subsequent inspection, and also can be deleterious for the operation of the parts, for example, because of the attraction of unwanted particles to the part by magnetic attraction [2].

10.1.1 INTRINSIC MAGNETIC PROPERTIES

In nondestructive evaluation, the magnetic measurements that depend on materials properties throughout the material are measurements of intrinsic properties. Whereas those that depend on local variations, such as leakage fields around cracks, are measurements of extrinsic properties. Among the intrinsic properties some are insensitive to material structure (such as saturation magnetization); whereas others are structure sensitive (such as coercivity). The structure sensitive properties are useful for nondestructive evaluation.

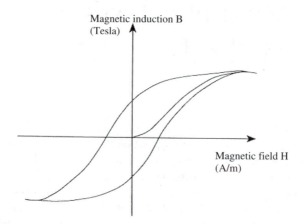

FIGURE 10.1 Variation of magnetization with magnetic field known as the hysteresis curve in which the magnetization does not completely recover its original value when a magnetic field is applied and then removed.

10.1.2 MAGNETIZATION CURVES AND HYSTERESIS

When a magnetic material is subjected to a cyclic magnetic field, the magnetization changes in a way that does not usually obey simplified linear equations. The magnetization changes in a nonlinear way with the field, and as the field is removed, the material retains some of its magnetization even at zero field. The variation of magnetization with field is usually plotted showing the magnetization along the y-axis and the magnetic field along the x-axis, and gives a characteristic magnetization curve for both increasing and decreasing magnetic fields known as the hysteresis curve [3]. An example of such a hysteresis curve is shown in Figure 10.1. The magnetization curve is stress dependent and, therefore, measurements of the magnetization curves can be used in the evaluation of stress [4].

10.1.3 DYNAMIC DEMAGNETIZING EFFECTS: REDUCING THE MAGNETIZATION TO ZERO

In many cases, it is desirable to demagnetize a test material prior to conducting any nondestructive testing on it, to put the test material in a known reproducible reference state [2]. To demagnetize a test material, it needs to be subjected to an externally applied magnetic field, which is then reversed and reduced successively at each cycle to arrive at the demagnetized state. Usually, a large number of cycles is required to perform this demagnetization, with a reduction in field amplitude of a few percent at each cycle. If the amplitude of the applied field at each successive cycle is reduced too quickly, then the magnetization can converge to a nonzero value, the magnitude of which depends on the rate of reduction per cycle and the number of cycles

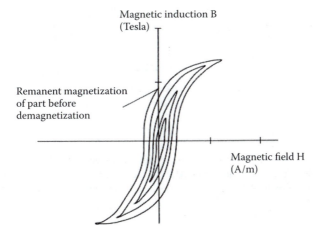

FIGURE 10.2 Decrease in amplitude of magnetization with time as a material is demagnetized.

used. Therefore, great care is needed to perform this demagnetization properly, and also, testing of the state of magnetization after the demagnetizing procedure is important [3]. The change in magnetization with time is cyclic with decaying amplitude, as shown in Figure 10.2.

10.1.4 STATIC DEMAGNETIZING EFFECTS: THE DEMAGNETIZING FACTOR

Even if the material is not subjected to a progressively reducing, cyclic magnetic field, the simple removal of the magnetic field to zero has the effect of reducing the magnetization, although the magnitude of this effect depends on the geometry.

When a material is magnetized it is in a high-energy state, and so there is naturally a tendency to demagnetize to reduce the magnetostatic energy associated with the material. This tendency can be explained in terms of the demagnetizing field, H_d, which is dependent on both the level of magnetization and the geometry of the test material. In other words, it is an extrinsic effect. The field inside a magnetic material of finite dimensions is smaller than what we would normally expect. The situation is shown in Figure 10.3.

The demagnetizing field is usually written as the product of the magnetization and the demagnetizing factor N_d:

$$H_d = N_d M. \tag{10.1}$$

The demagnetizing field is therefore (1) proportional to the level of magnetization, (2) opposite in direction to M, and (3) dependent on sample geometry.

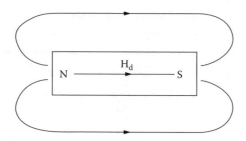

FIGURE 10.3 Contributions to the magnetic field in the vicinity of a magnetic dipole. The applied field H_a causes a magnetization M, which can be represented as a magnetic dipole. This dipole sets up a field H_d acting in opposition to the magnetic field which caused the magnetization.

If there is an applied magnetic field H_a, then the total magnetic field is the difference between the applied and the demagnetizing fields,

$$H = H_a - H_d. \tag{10.2}$$

Sometimes we wish to express the fields in terms of B and H

$$M = \chi H \tag{10.3}$$

$$H = H_a - N_d \chi H \tag{10.4}$$

$$H = \frac{H_a}{1 + N_d \chi} = \frac{H_a}{1 + N_d (\mu_r - 1)}, \tag{10.5}$$

and because $B = \mu_o(H + M)$

$$H = H_a - N_d \left(\frac{B}{\mu_o} - H \right) \tag{10.6}$$

$$H(1 - N_d) = H_a - N_d \frac{B}{\mu_o} \tag{10.7}$$

$$H = \frac{H_a - N_d \dfrac{B}{\mu_o}}{(1 - N_d)}. \tag{10.8}$$

FIGURE 10.4 Schematic of a material with regions of different relative permeabilities μ_1 and μ_2.

All of these equations demonstrate that the actual magnetic field, H, is different from the applied magnetic field in a way that depends on both the level of magnetization, M, and the geometry, N_d.

10.1.5 EQUATIONS GOVERNING THE DEMAGNETIZING EFFECTS IN INHOMOGENEOUS MATERIALS

In any situation where we have magnetic inhomogeneity as a result of differences in permeability, the magnetic field will be altered locally. Consider the situation depicted in Figure 10.4.

In this case, the magnetic field in region μ_1 will be the applied field minus the demagnetizing field,

$$H = H_a - N_d \left(\frac{\mu_1}{\mu_2} - 1 \right) H. \tag{10.9}$$

So, for example, in a particular case if region 1 is iron ($\mu_1 = \mu_{iron}$) and region 2 is air ($\mu_2 = 1$), then

$$H = H_a - N_d (\mu_{iron} - 1) H \tag{10.10}$$

$$H = H_a - N_d \chi_{iron} H = H_a - N_d M. \tag{10.11}$$

This is the known result that was given previously. The general equation for the field, taking into account the corrections for demagnetizing fields, can be written in several different but equivalent forms, including the following form, which is useful for considering the magnetic field in region 1 of an inhomogeneous material,

$$H = \frac{H_a}{1 + N_d \left(\dfrac{\mu_1}{\mu_2} - 1 \right)} = \frac{\mu_2 H_a}{\mu_2 + N_d (\mu_1 - \mu_2)}. \tag{10.12}$$

So that if $\mu_1 = \mu_2$, meaning no magnetic interface, then $H = H_a$.

If $\mu_1 > \mu_2$, which applies for example to a magnetic material of finite dimensions in free space, then $H < H_a$.

If $\mu_1 < \mu_2$, which applies for example to a crack in a magnetic material, then $H > H_a$.

WORKED EXAMPLE

Calculate the magnetic field inside a material with $N_d = 0.02$ at an applied field of $H_a = 80$ kA/m and a magnetic flux density of 0.9 T.

Solution

$$H = \frac{H_a - N_d \dfrac{B}{\mu_o}}{(1 - N_d)} \tag{10.13}$$

$$= 1.02\ (80{,}000 - (0.02)(0.9)/12.56 \times 10^{-7})$$

$$= 66.98 \text{ kA/m}$$

10.2 MAGNETIC METHODS FOR EVALUATION OF DEFECTS

10.2.1 INTRINSIC METHODS

There are several magnetic methods for NDE that depend on the measurement of the inherent properties of the magnetic material. These include methods dependent on the measurement of magnetization curves or hysteresis curves, Barkhausen effect, magnetoacoustic emission, and magnetomechanical effects. Each of these properties is sensitive to the structure of the material, such as the grain size, grain orientation (texture), presence of second phase material such as carbides in steels, applied and residual stress, build-up of dislocations, etc. Once a relationship has been found between the property of interest for materials evaluation (such as residual stress) and the magnetic measurement property (such as coercivity), the magnetic measurement can be used for evaluation of the condition of the material. The relationship can be an empirical one, such as a calibration curve based on an assessment of practical measurements, or a theoretical one based on models of materials behavior.

10.2.2 EXTRINSIC METHODS

Extrinsic methods are concerned with the identification of flaws or defects in the material. These methods include the use of magnetic particle inspection (MPI), magnetic flux leakage (MFL), and magnetic imaging based on flux leakage images or magnetic force microscopy.

10.2.3 DETECTION OF FLAWS AND CRACKS USING MAGNETIC FLUX LEAKAGE

When the direction of magnetic flux in a material is perpendicular to a flaw, the flaw causes the magnetic flux to be diverted because it represents a region of low permeability surrounded by regions of good material of high permeability. This is completely analogous to the situation described in Section 10.1.5, where the diversion of the flux is considered to be the result of a demagnetizing field. This method of stray flux detection has become one of the most popular methods for detection of defects in ferrous materials [5]. The effect is greatest in those situations where the flaw is perpendicular to the flux, and the effect decreases as the angle is reduced. When the flaw is parallel to the field, it is difficult, if not impossible, to detect the flaw by this method. Both surface flaws and subsurface flaws that are close to the surface are detectable by this method.

There are a variety of ways to produce a magnetic field across a test material. One way is to use an electromagnet in the form of a high-permeability core wrapped with a field coil that can be moved across the surface of the test material. The electromagnet will produce a dipole field, as shown in Figure 10.5, and these flux lines are disturbed by the presence of a flaw.

Another way to produce a magnetic field is the "threaded tube" method, in which an electrical conductor carrying a current is passed through the test material. This generates a circulating field, which can be used to detect the presence of axial flaws as shown in Figure 10.6. The strength of the magnetic field

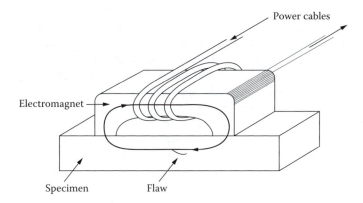

FIGURE 10.5 Schematic of an electromagnet on the surface of a test material showing the distribution of flux lines.

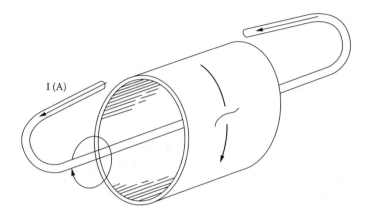

I (A)

FIGURE 10.6 The threaded tube method of producing a circulating field in the test specimen.

generated in the wall of the tube is largely independent of the position of the conductor within the tube [9].

According to Stumm [5], magnetic flux leakage measurements can detect flaws in ferrous tubes or pipes with sizes in the range of 5 to 10% of wall thickness.

10.2.4 DESCRIPTION OF CRACKS AS MAGNETIC DIPOLES

Magnetic dipoles produce a field pattern that has a distinctive form and which can be described by analytical equations that can be found in most standard textbooks. The presence of planar cracks in materials can also be described approximately as an assembly of dipoles, leading to dipole-like leakage fields in their vicinity. For many purposes this is an adequate description of such fields, although if exact descriptions are required, a more sophisticated approach using finite element modeling is required. The variation of the tangential component of leakage flux with position across a planar crack is shown in Figure 10.7. It can be seen from this that the tangential component is unipolar. In contrast, the normal component of the flux density across a planar crack is bipolar (not shown).

The dipole model for the leakage flux density around a crack can give a reasonably good description of the crack, as in the work of Zatsepin and Schcherbinin, which is shown in Figure 10.8. In this case, the field calculations were performed assuming only that the crack could be represented as an assembly of dipoles.

10.2.5 EQUATIONS FOR FIELDS AROUND CRACKS

Starting from the demagnetizing field due to a simple dipole as an approximation to the field around the flaw, the actual field H can be calculated in terms of the applied field H_a using equation (10.2) and equation (10.12) for fields around

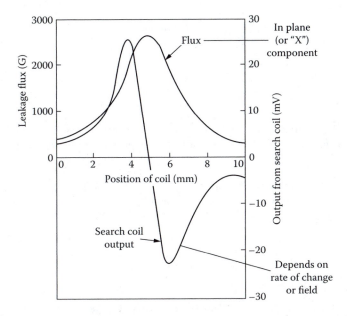

FIGURE 10.7 Variation of the "in plane" component of leakage flux in the vicinity of a planar crack and the output of a search coil scanned across the crack at constant speed.

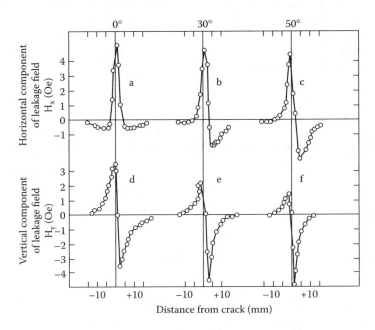

FIGURE 10.8 Calculated leakage field due to a flat-bottomed flaw using the dipole model; a, b, and c are the tangential components of field, and d, e, and f are the normal components after Zatsepin and Schcher binin.

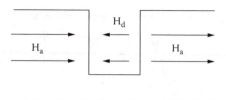

FIGURE 10.9 A simplified rectangular crack in a magnetic material gives rise to an additional field, H_d, as a result of the magnetic poles that appear on the surface of the crack.

inhomogeneities in magnetic materials, if region 1 is the flaw and region 2 is the magnetic material, this will give the following

$$H = \frac{H_a}{1 + N_d \left(\dfrac{\mu_{flaw}}{\mu_{iron}} - 1 \right)}, \tag{10.14}$$

and when $\mu_{flaw} < \mu_{iron}$, then the field H in the crack will be greater than the applied field, H_a. It is more difficult to drive magnetic flux across the crack because of the low relative permeability of a crack compared with the rest of the material. However, we can think of the problem in terms of a region of missing magnetic material in an otherwise continuous material medium, as depicted in Figure 10.9.

The strength of the demagnetizing field depends on the difference in magnetization between the crack (presumably air) and the magnetic material (usually iron or steel), according to equation (10.14).

WORKED EXAMPLE

Calculate the field in the air gap in Fig. 10.9, where $H_a = 80$ kA/m and $N_d = 0.2$ for the flaw. The relative permeability of the material on either side of the flaw is $\mu_r = 8.96$, and for the flaw itself $\mu_r = 1$.

Solution

$$H = \frac{H_a}{1 + N_d \left(\dfrac{\mu_{flaw}}{\mu_{iron}} - 1 \right)} \tag{10.15}$$

$$H = \frac{80,000}{1 + 0.2 \left(\dfrac{1}{8.96} - 1 \right)} = 97.3 \text{ kA/m} \tag{10.16}$$

10.2.6 EXAMPLES OF LEAKAGE FIELD CALCULATIONS

In almost all situations, except for very simplified cases, the leakage fields due to flaws are too complicated for simple analytical formulae, such as those obtained using demagnetizing field calculations. In the more realistic and complicated situations, the leakage field needs to be calculated using numerical methods, such as finite element modeling.

Examples of leakage fields calculated in the vicinity of three different types of flat-bottomed flaws from the work of Lord are shown in Figure 10.10.

In practical inspection situations, such as the use of magnetic flux leakage to inspect pipelines for defects, the magnetic flux leakage signal can be affected by a number of factors, including the velocity of the sensor and the magnetic properties of the material, which can vary from location to location. Such information may be very difficult, or even impossible, to obtain and therefore interpretation of results can be problematic. However, use of advanced signal processing techniques to compensate or minimize these effects is possible, as described by Mandayam et al. [6].

Magnetic flux leakage signals can be used to create an image of the magnetic field on the surface of a test specimen [7]. This method, termed *magnetography*, can be used to identify regions of a material that have flaws or regions of high stress. Magnetography can also provide a useful quantitative archival record of the condition of the material, but it takes longer to complete than the simple magnetic particle inspection.

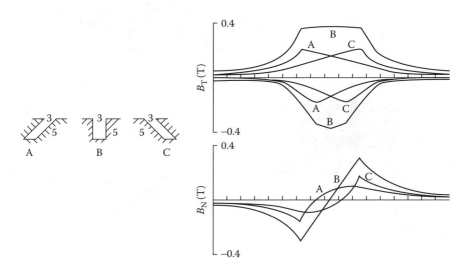

FIGURE 10.10 Calculated leakage field profiles for the tangential and normal components of magnetic field due to flaws in a semi-infinite magnetic medium.

10.3 MAGNETIC PARTICLE INSPECTION

10.3.1 VARIOUS PROCEDURES FOR GENERATING THE MAGNETIC FIELD FOR MPI

There are a variety of different procedures for carrying out magnetic particle inspection (MPI) [8]. Most of these relate to the way in which the magnetic field is generated in the material. The basic guidelines for satisfactory magnetic particle inspection have been described by Betz [9] and King [10]. There are also some variations on the magnetic particles that are used to detect the flaws; the main choice being between using dry powder or wet ink methods. Different types of magnetic powders or inks are in current usage, including aerosol-based magnetic powders [11]. A variety of methods for producing the field are shown in Figure 10.11.

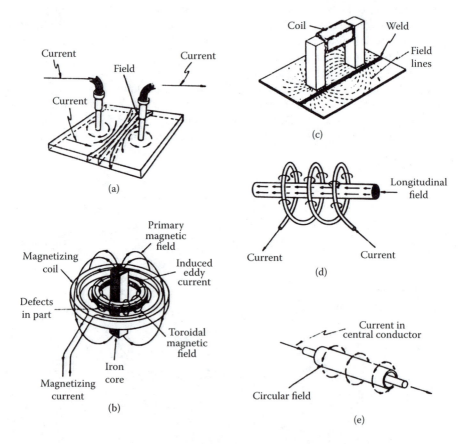

FIGURE 10.11 Various methods for creating the magnetic field in a test material (a) current prods, (b) AC threaded tube method with current circulating tube, (c) electromagnetic core, (d) encircling coil (figure shows a coil wound directly on the sample but a rigid coil can also be used), and (e) standard threaded tube method for either AC or DC currents, with magnetic field circulating tube, for axial cracks.

In most cases, care has to be taken to get the correct field strength in the test specimen, because the magnetic field varies rapidly with position. There is an exception in the case of the threaded bar method, in which it has been shown by Edwards and Palmer [12] that the field strength is largely independent of the location of the conductor within the tube. Also, the amount of material inside the magnetic inspection coil — the so-called "fill factor" — affects the results of the inspection [13]. Essentially, this depends on the relative cross-sectional areas of the coil and the test specimen, and defines the number of ampere-turns (or magnetomotive force) needed to get optimum results.

10.3.2 Practical Considerations for Use in MPI

There are very few exact formulae for use in quantitative MPI. This makes the successful use of MPI critically dependent on the establishment of inspection procedures and the experience of the operator [14]. One necessity that is generally acknowledged is that a flux density of 1.1 T is considered to be sufficient for MPI of flaws. The field can be produced by a permanent magnet or by an electromagnet.

Procedures for MPI have been automated, and the introduction of signal processing for image analysis has led to significant improvements in the method [15].

The problem of the optimum level of magnetization for detection of flaws using MPI is one that recurs frequently, probably because there is no completely general optimum when the inspection is dependent on so many competing factors. The optimum level of a magnetization depends on the size of defects, the depth below the surface, the shape of a defect and its orientation, and the magnetic characteristics of the test material. Massa [16] conducted tests and reported a series of values of optimum magnetization depending on geometry of the defect for a series of standard (i.e., simplified) defects.

10.3.3 Rigid Coils

Broadly, there are two types of coil that are used. One of these is a preformed coil, known as a rigid coil. This comes already wound with a predetermined geometry and a fixed number of turns. It is often used with the test material placed inside it, so that it surrounds the region of the test material to be inspected.

To determine the current needed in a rigid coil, the following empirical equation is used for a cylindrical test material,

$$Ni = k\frac{D}{\ell}, \tag{10.17}$$

where N is the number of turns on the coil, i is the current in amperes passing through the coil, D is the diameter of the test material in meters, and l is the length of the test material. k is a constant that has the value of 32,000 for DC currents, 22,000 for either AC or full-wave rectified currents, and 11,000 for half-wave rectified currents. The above equation can be used reliably for test materials that have a length-to-diameter ratio in the range $5 < l/D < 20$, and which occupy less than 10% of the cross-sectional area of the coil.

10.3.4 FLEXIBLE COILS

A second type of coil that is widely used is a coil that is wound directly on to the test material. This is known as a "flexible coil." The current required for inspection using these types of coils can be calculated from empirical equations, depending on whether the current is DC or AC.

For a DC current source

$$i = 7.5\left(t + \frac{d^2}{4t}\right), \tag{10.18}$$

and for an AC current source

$$i = 7.5\left(10 + \frac{d^2}{40}\right). \tag{10.19}$$

In these equations, i is the peak value of current in amps, t is the wall thickness of the test material (or the radius if it is a solid bar) in millimeters, and d is the spacing distance in millimeters between the windings.

10.3.5 CURRENT NEEDED TO MAGNETIZE STEELS

In cases where magnetization is achieved through the flow of current through the test material, there are also empirical formulae for the appropriate levels of current. As shown in Figure 10.12, if a current, i, is passed axially along a rod of magnetic material of diameter d, then the current required for MPI is dependent on whether the current is DC or AC, and whether the cross-section of the rod is equiaxed or not.

Shape	DC	AC
Equiaxed	28 A/mm of diameter	20 A/mm of diameter
Nonequiaxed	9 A/mm of periphery	6 A/mm of periphery

10.3.6 KETOS RING TEST

This consists of a ring of magnetic material in which identically sized holes have been drilled at predetermined distances from the outer circumference. Current is

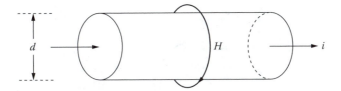

FIGURE 10.12 Configuration for magnetic particle inspection using current flow in the test material.

passed though a conductor that passes through the center of the ring perpendicular to the plane of the ring. Magnetic particles are dusted onto the outer circumferential surface of the ring, and recommended numbers of detectable holes are given for various values of the electrical current passing through the center of the ring. This is used for investigating what current is needed for certain powders to detect flaws, or what powders are needed to detect flaws with a given current. Figure 10.13 shows a Ketos ring with the configuration of holes at different distances from the circumference, and the specification for the distances of these holes from the outside surface. The holes are 1.75 mm in diameter, and their distances from the surface vary from 1.8 to 21.3 mm. The holes are spaced at circumferential distances of 19 mm.

10.3.7 PIE GAUGE TEST

The pie gauge is a high-permeability polygon with six or eight sectors. There are nonmagnetic gaps of fixed width in the magnetic material, which can be used to

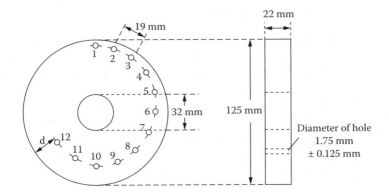

FIGURE 10.13 The geometrical configuration of the Ketos ring which is used for calibrations for magnetic particle inspections.

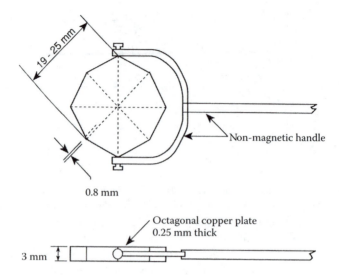

FIGURE 10.14 A pie gauge consisting of a polygonal plate with several magnetic sections, separated by nonmagnetic gaps that can be used to determine the approximate direction of a magnetic field at the surface of a magnetic test material.

find the approximate orientation of the magnetic field on the surface of a test material. In practice, the pie gauge is placed on the surface of the test material, and magnetic powder is dusted on to the surface of the pie gauge. The field at the surface of the test material will lie nearly perpendicular to one of the gaps, which accumulates the most magnetic powder. Figure 10.14 shows a top and side view of the pie gauge.

10.3.8 SPECIAL TECHNIQUES FOR MPI

There is a variety of special techniques that can be used in connection with magnetic particle inspection and which provide permanent records of the materials tested. These include the use of magnetic rubber, magnetic tape, magnetic printing, magnetic painting, and pressure sensitive tape.

Magnetic rubber comprises a suspension of magnetic particles in liquid rubber, which is then poured on to the test material. The material is magnetized, and the rubber is allowed to cure. The solid rubber then has a magnetic pattern that matches the pattern at the surface of the test material. This imprint can then be used to study the location of flaws in the material in much the same way as a conventional magnetic particle inspection. The difference is that the magnetic rubber gives a more permanent record of the state of the material.

Magnetic tape can be used in many situations, particularly those inspections in which the conventional powder MPI cannot be used, such as inspections that need to be conducted under water. The magnetic tape, which consists of fine magnetic particles contained in a polymeric binder, is attached to the surface of the test material. A field is applied to magnetize the part, and the field pattern at the surface of the test material is detected by the magnetic tape, resulting in a magnetic pattern on the tape that matches the pattern at the surface of the test material. The tape can then be removed and provides a permanent record of the magnetic flux pattern at the surface of the test material. The magnetic pattern on the tape can then be examined under well-controlled laboratory conditions using magnetometers.

Magnetic printing is the name given to a technique in which the test material is sprayed with a uniform monochromatic plastic coating, which is allowed to dry. Then a magnetic field is applied, and magnetic particles are dusted onto the surface in the same way as for conventional MPI inspection. The excess particles are removed, leaving an MPI pattern showing the location of flaws. A clear plastic coating is then applied to fix the magnetic particles in place. This is allowed to dry, and the dried film is then removed to give a permanent record of the magnetic pattern at the surface of the test material.

Magnetic painting, also known as magnetic ink, is an alternative technique to the dry powder method that works well in many situations. The "ink," which consists of a suspension of magnetic particles in a liquid carrier, is brushed onto the surface of the test material. A magnetic field is applied, and the magnetic particles accumulate preferentially at locations with a high magnetic field gradient. This results in dark lines at the locations of flaws, but elsewhere a grey background. The contrast is often superior to the dry powder MPI method.

Pressure-sensitive tape is sometimes used in combination with conventional MPI. After the MPI pattern has been produced on the surface of the test material in the usual way, pressure-sensitive adhesive tape is applied to the surface to fix the magnetic particles in position. The tape is then removed with the magnetic particles attached, which thereby provides a permanent record of the magnetic pattern.

EXERCISES: MAGNETIC TESTING METHODS

10.1 Explain the ideas behind the MPI method as it is used in nondestructive testing. To obtain the required number of ampere-turns to magnetize a steel part of length 0.5 m and diameter 75 mm for MPI, a half-wave rectified current source is to be used. If there are 300 turns in a preformed rigid coil that is being used to magnetize the specimen, calculate the value of current that is needed.

10.2 Explain the ideas behind the magnetic flux leakage method for nondestructive testing. For magnetic crack detection of circumferential defects in the circumferential welds of a 6-in. diameter pipe, a flexible coil of insulated wire is wound around the outside of the pipe so that two adjacent turns of the coil are on each side of the weld. The cable can carry a current of 1000 A AC (peak value). What should the maximum spacing between the coil turns be?

10.3 If a leakage flux density of 1.1 T is needed to get an adequate accumulation of magnetic particles for an indication of surface flaws using MPI, determine the level of applied field, H, in amps per meter, required to obtain such an indication on a particular piece of steel with relative permeability 100 and a surface flaw with an aspect ratio giving a geometrical factor $N_d = 0.5$. If this field is produced by a long solenoid with 25 turns/mm, calculate the current needed in the coil.

REFERENCES

1. Jiles, D.C., Review of magnetic methods for NDE, *NDT Int* 21, 211, 1988 and 23, 83, 1990.
2. Nippes, P.I. and Galano, E.N., The need for integrating DC downcycle demagnetization into magnetic particle inspection, *Mater Eval* 58, 365, 2000.
3. P. I. Nippes and E.N. Galano, How and why to measure magnetism accurately, *Mater Eval* 60, 507, April 2002.
4. Langman, R., Measurement of mechanical stress in mild steel by means of rotation of magnetic field, *NDT Int* 14, 255, 1981 and 15, 91, 1982.
5. Stumm, W., Magnetic stray-flux measurement for testing welded tubes on line, *NDT Int* 9, 3, 1976.
6. Mandayam, S., Udpa, L., Udpa, S.S., and Lord, W., Signal processing for in-line inspection of gas transmission pipelines, *Research in NDE*, 8, 233, 1996.
7. Forster, F., Developments in the magnetography of tubes and tub welds, *Nondestr Test Eval Int* 8, 304, 1975.
8. Bailey, W.H., Magnetic particle inspection, *Mater Eval* 42, 962, 1984.
9. Betz, C.E., *Principles of Magnetic Particle Inspection*, Magnaflux Corp., Chicago, IL, 1967.
10. King, W.G., A practical introduction to magnetic particle testing: basic rules for satisfactory detection, *Nondestr Test Eval Int* 1, 84, 1967.
11. Shreve, D.A. and Cheddister, W.C., The evolution of magnetic particles follows industry needs, *Mater Eval* 53, 883, 1995.
12. Edwards, C. and Palmer, S.B., An analysis of the threaded bar method of magnetic particle flaw detection, *NDT Int* 14, 177, 1981.
13. Dunckley, B.L., The intermediate fill factor equation as used in longitudinal magnetization for magnetic particle testing, *Mater Eval* 58, 1361, December 2000.
14. Kleven, S., Pieko, E.J., and Hughes, T., A critical commentary on magnetic particle inspection revisited, *Mater Eval* 55, 23, January 1997.
15. Borucki, J.S., Development of automated magnetic particle testing systems, *Mater Eval* 49, 324, March 1991.
16. Massa, G.M., Finding the optimum conditions for weld testing by magnetic particles, *Nondestr Test Eval Int* 9, 16, 1976.

FURTHER READING

Betz, C.E., *Principles of Magnetic Particle Inspection*, Magnaflux Corp., Chicago, IL, 1967.

Blitz, J., *Electrical and Magnetic Methods of Nondestructive Testing*, 2nd ed., Chapman and Hall, London, 1997, 261 pp.

http://www.ndt-ed.org/EducationResources/CommunityCollege/MagParticle/cc_mpi_index.htm.

Lo, C.C.H., A review of the Barkhausen effect and its applications to nondestructive testing, *Mater Eval* 62, 743, July 2004.

Schmidt, J.T. and Skeie, K., *Handbook of Nondestructive Testing: Magnetic Particle Testing*, Vol. 6, ASNT, Columbus, 19810.

Zatsepin, N.N. and Schcherbinin, V.E., *Sov. J. NDT*, 2, 50, 1966.

11 Radiographic Testing Methods

X-ray inspection methods are among the oldest nondestructive inspection methods and are still some of the most widely used. X-ray diffraction can be used for determination of crystal structure, texture, and lattice parameters, and x-ray radiographic methods are used for locating inhomogeneities in materials. This chapter discusses the factors that affect the formation of an x-ray image, including geometrical set-up, beam divergence, properties of the x-ray film, contributions to "unsharpness" of the image, and how these various conditions for radiography can be controlled for best results in the final radiograph.

11.1 X-RAY IMAGING

X-rays can be used to form images, usually with photographic film, in much the same way as visible light is used to form images, with the difference of course that x-rays can penetrate materials, and therefore give an indication of the internal structure [1]. Gamma rays, which originate from radioactive materials but are otherwise a type of electromagnetic radiation similar to x-rays, can also be used to form images [2].

11.1.1 X-RAY IMAGES

It is well known that x-ray images are not particularly sharp when compared, for example, with photographic images. This has nothing to do with diffraction of the x-rays which, being of high energy compared with optical photons, suffer much less diffraction. This phenomenon is also not just due to the coarseness of the grains in the x-ray film, but has to do instead with several factors, some of which are geometrical in nature and would therefore arise even when other detection techniques are used, and some of which are related to the quality of the detector or the film.

Detection of x-rays can be achieved in a number of ways. The traditional way is through the use of x-ray film; however, electronic instrumentation can be used for real time detection. This includes the use of gas-filled and solid state detectors [3], in which the detection of incident x-rays generates a voltage, which by scanning over an area can give an image of the spatial variation of x-ray intensity.

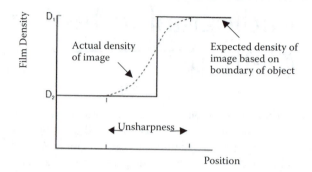

FIGURE 11.1 Geometrical considerations affecting "unsharpness" of an x-ray radiographic image.

11.1.2 BLURRING OF IMAGES

We consider an extended source of x-rays with spatial extent F, at a distance L_o from an object, with the object at a distance l from the film. This will cause blurring of the image at the film (Figure 11.1) and is known technically as "unsharpness," which can be calculated from the geometry. Such blurring of the image therefore cannot be completely avoided.

Thus, although the boundaries of the object that is being x-rayed may be sharp, the image of these boundaries will be blurred on the radiograph. The unsharpness is then a measure of the length of the transition region on a radiograph that corresponds to a completely sharp (step change) boundary in the test object. The unsharpness is reduced when the object is moved closer to the film, and when the spatial extent of the x-ray source is reduced.

11.1.3 PROJECTION RADIOGRAPHY

In the above example, the object under test and the radiographic film did not touch. This configuration is known as projection radiography, as distinct from contact radiography, when the object is in contact with the film. When taking a radiograph, this results in a dark (fully-shaded) region and a semidark (partially-shaded) region. As shown in Figure 11.2, the completely dark region of the image, known as the umbra, has a finite spatial extent S, whereas the region of partial darkness, known as the penumbra, has a spatial extent P. The extent of the penumbra shows the inability to clearly define the boundary of an object in the image, and is a measure of unsharpness. This increases the further the object is from the film.

11.1.4 GEOMETRICAL UNSHARPNESS

The unsharpness is therefore also the minimum separation distance that can be resolved between two features in the test object under particular inspection

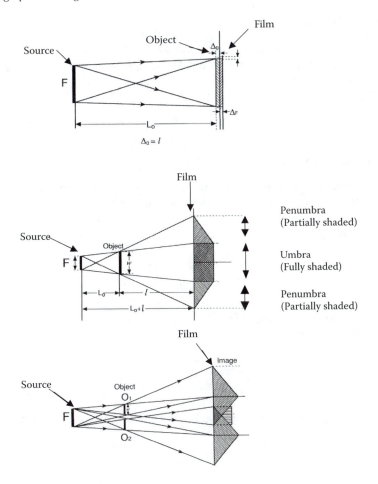

FIGURE 11.2 Setup for projection radiography in which a fully-shaded region and a partially- shaded region are produced, depending on the source-to-object and object-to-film distances.

conditions. It can be calculated from the following equation [4, p.117]:

$$U_g = F \frac{l}{L_o}, \tag{11.1}$$

where F is the width of the x-ray source, l is the object-to-film distance, and L_o is the source-to-test object distance.

11.1.5 MAGNIFICATION

When projection radiography is used, the size of the image will be larger than the object, and the magnification factor, M_G, can be found from the geometry,

$$M_G = \frac{L_o +}{L_o}.$$ (11.2)

11.1.6 OTHER GEOMETRICAL FEATURES

The properties of the image that are of interest are the size of the umbra and the size of the penumbra. The limit of detection indicates the smallest feature that can be resolved in the image.

Size of umbra S:

$$S = \frac{(L_o + \ell)w - F\ell}{L_o}$$ (11.3)

where w is the width of the object.

Size of penumbra P:

$$P = \frac{F}{L_o}$$ (11.4)

Limit of detection W: (This is the condition when the size of the umbra reaches zero.)

$$W' = \frac{F\ell}{L_o + \ell}$$ (11.5)

11.1.7 CONTACT RADIOGRAPHY

The unsharpness of the radiographic image is minimized if the test object is in contact with the radiographic film. In principle, this corresponds to $l = 0$ in the above equations, but in practice there is always a finite distance from the test object to the film, which can often be approximated as the thickness of the test object. In addition, the thickness of the radiographic film should be taken into account.

If the film thickness is Δ_F and the object thickness is Δ_O, then the geometrical unsharpness for contact radiography is

$$U_g = \frac{F}{L_o}(\Delta_O + \Delta_F),$$ (11.6)

which is nonzero.

11.2 RADIOGRAPHIC FILM

11.2.1 Properties of the X-Ray Film

The radiographic film that is used to produce radiographs consists of silver bromide particles in an organic matrix. The silver bromide content typically amounts to a few mg per cm^2 of cross-sectional area of the film. Only the AgBr absorbs the x-ray to any significant extent, and about 5% of the energy absorbed by the silver bromide results in precipitation of silver atoms to form the image.

Film unsharpness (as distinct from geometrical unsharpness) is caused by secondary electrons emitted from the silver bromide grains, which affect neighboring grains. As a result of this, the film unsharpness increases with the energy of the incident x-rays.

Comparison and classification of x-ray films, on the basis of film granularity and signal-to-noise ratio, have been, until recently, mostly qualitative. However, quantitative comparisons of films can be made on the basis of their "image quality index" (IQI) [5]. The parameters of interest are equivalent penetrameter sensitivity (EPS), grain size (d), and "crack detection index" (CDI), where CDI $= 20 \log(1/l_{min})$, and l_{min} is the smallest crack size that can give a visible image on the film.

11.2.2 Film Exposure

The density of the radiograph is determined by the exposure to radiation ε:

$$\varepsilon = I\,t, \tag{11.7}$$

where I is the intensity of the radiation, and t is the time of exposure. The film density changes with exposure. Normally, all of the film is exposed for the same amount of time, so that the variation of exposure over the area of the film varies with the intensity of incident radiation over the same area:

$$\varepsilon(x,y) = I(x,y)t. \tag{11.8}$$

11.2.3 Photographic Density

The density of a radiograph changes from one location to another across the film. This density is usually detected by the use of back-lighting, so that the film density is determined by the intensity of transmitted light. Consider a final radiographic film, which is illuminated from behind to show the flaws. These could be regions of high photographic density on the film. If light is being transmitted through the radiograph, then we could say that the brightness of the light that has passed through the radiograph, B_T, is related to the brightness of the light that was incident on the radiograph, B_I, according to the relation

$$B_T = B_I 10^{-D}, \tag{11.9}$$

or equivalently

$$D = \log_{10}\left(\frac{B_I}{B_T}\right),$$ (11.10)

where D is the density of the film.

11.2.4 Radiographic Contrast

There is a quantitative definition of contrast that is used for images on films. This is the difference in density between two regions of the film,

$$C_s = D_1 - D_2.$$ (11.11)

This difference in density will depend on the exposure of the film at the different locations being compared.

Any radiographic image will suffer from blurring (also known technically as unsharpness), and this represents a loss of information between the spatial variation of transmitted x-ray intensity, and the recording of that variation in intensity on a film. It limits the ability to distinguish detail from background noise, and also limits ability to resolve features. A "resolving power function," which expresses the ability to detect changes in film density, has been described by Notea [6] so that a change in film density from one location to the next can be detected when it exceeds the random fluctuations (noise) in the film. This quantitative approach allows optimization of radiographic system design.

11.2.5 Film Density and Gradient

If we therefore compare two regions with radiographic densities D_1 and D_2, then assuming that the incident light intensities on the two regions are the same, the radiographic contrast, C_s, between the two regions can be defined in terms of the transmitted intensities B_{T1} and B_{T2} in the two regions, or in terms of the exposures ε_1 and ε_2,

$$C_s = G \log_{10}\left(\frac{B_{T2}}{B_{T1}}\right) = G \log_{10}\left(\frac{\varepsilon_2}{\varepsilon_1}\right),$$ (11.12)

where the term G is a characteristic of the film, known as the film gradient. Typically, this has a value of $G = 4$. Over the linear region of the density vs. log exposure curve, the film gradient, G, which measures how the film contrast varies with exposure, [7] is given by

$$G = \frac{D_1 - D_2}{\log_{10}\varepsilon_1 - \log_{10}\varepsilon_2} = \frac{D_1 - D_2}{\log_{10}\left(\frac{\varepsilon_1}{\varepsilon_2}\right)}. \tag{11.13}$$

Normally the exposure time is the same at all locations on the radiograph, so the ratio of exposures reduces to the ratio of intensities of radiation at the two locations being compared. Then $\frac{\varepsilon_1}{\varepsilon_2} = \frac{I_1}{I_2}$, and therefore

$$G = \frac{D_1 - D_2}{\log_{10}\left(\frac{I_1}{I_2}\right)}. \tag{11.14}$$

Hence, we can relate contrast between two regions of the film to (1) the film gradient and (2) intensities of radiation incident on the film:

$$C_s = G \log_{10}\left(\frac{I_1}{I_2}\right). \tag{11.15}$$

11.2.6 UNSHARPNESS

For a given radiographic film, the film unsharpness varies with energy of the incident x-rays, and this variation is typically like that shown Figure 11.3. When making a

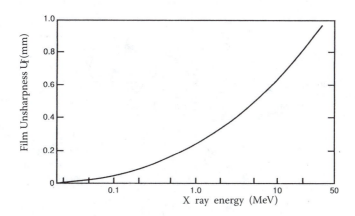

FIGURE 11.3 Variation of film unsharpness, U_f, with x-ray generator voltage.

radiograph, the optimum conditions are obtained if the geometrical unsharpness equals the film unsharpness.

11.3 RADIOGRAPHS

Different chemical species within a sample result in particular paths for x-rays being more or less attenuating than others. The absence of material also reduces attenuation (see Figure 11.4). Therefore, radiographs can be produced with different intensities at different locations on the radiograph, depending on how much and what kind of material was in the path between the radiation source (normally an x-ray set) and the detector (normally a radiographic film).

11.3.1 BEAM DIVERGENCE

In the case of a point source of radiation, the intensity decreases inversely with the square of the radial distance, r, from the source, as described earlier. This is the $1/r^2$ law and is well-known. When the beam of radiation diverges, and the

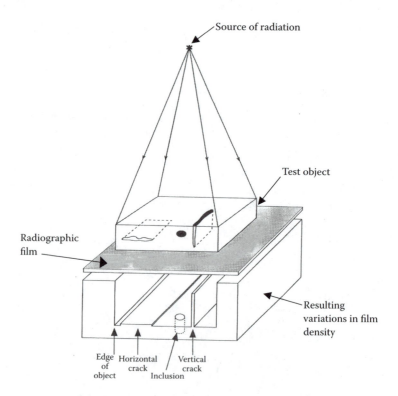

FIGURE 11.4 Schematic showing how a radiograph represents the presence of flaws in a material as variations in intensity on a film.

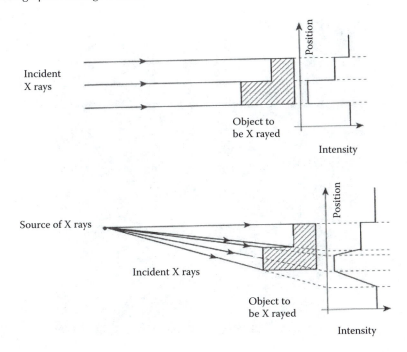

FIGURE 11.5 Examples of how transmitted intensity of radiation (and hence density on a radiograph) varies with position.

test material has a finite width, then there will be some shadowing on the radiographs due to the fact that the beam has passed through different thicknesses of material in order to reach the detector (film). This is shown in Figure 11.5. In one case, a parallel beam is used, giving sharp features of either light or dark (umbra or shaded) on the radiograph. In the other case, a divergent beam is used, giving regions of light, dark, and intermediate (partially-shaded or penumbra).

11.3.2 Examples of How Material Defects Can Appear in a Radiograph

Of course, now with the possibility of digital radiographs, either directly through electronic detection of x-rays, or by post-processing of traditional x-ray films to give digital images, a number of image processing options become available for improved analysis of data [8] (see Figure 11.6 to Figure 11.8).

Real-time imaging of x-rays, as is achieved in real-time microfocus x-ray systems [9], can have significant advantages over conventional detection because the process can be faster than the use of x-ray films, and can be used as a component of real- time process control. This has advantages in terms of time and cost on nondestructive inspection. In addition, the results of such digital real-time microfocus x-ray systems can be subjected to image analysis.

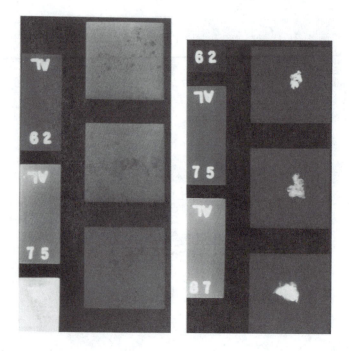

FIGURE 11.6 How inclusions appear in a radiograph.

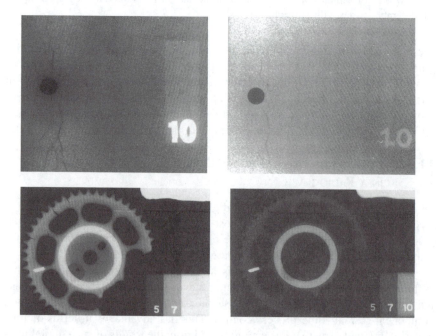

FIGURE 11.7 How cracks appear in a radiograph.

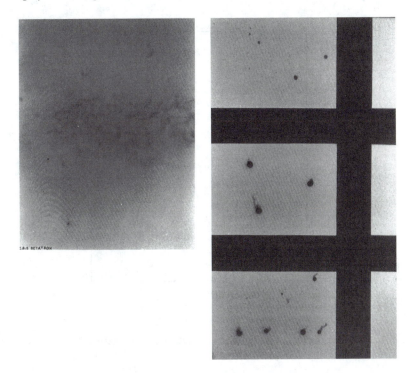

FIGURE 11.8 How holes appear in a radiograph.

Automated identification of various different types of defects detected in radiographs is now a well-developed technique. Computer-aided identification, including background subtraction and histograph thresholding, are used to separate defect signals from background noise on the radiographic image [10].

11.3.3 OPTIMUM EXPOSURE OF RADIOGRAPHS

It is clear that both underexposed and overexposed radiographs are poor for identifying flaws. Therefore, there must be an optimum exposure for radiographs, and we need to find out how to determine this optimum. A fairly obvious way to do this is by maximizing the contrast of the radiograph.

11.3.3.1 Variation of Radiographic Density with Exposure

The variation of radiographic film density with exposure usually follows a relationship of the type shown in Figure 11.9. Over part of the range of exposure, the rate of change of density is linear with logarithm of exposure, and the slope of this portion of the curve is known as the film gradient G [7].

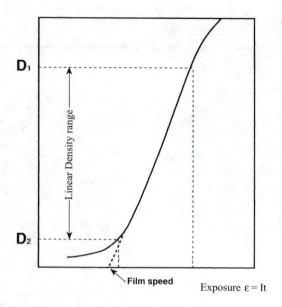

FIGURE 11.9 Dependence of radiographic film density on logarithm of exposure.

The radiographic contrast, C_s, is then

$$C_s = D_1 - D_2 = G(\log \varepsilon_1 - \log \varepsilon_2)$$

$$= G \log\left(\frac{\varepsilon_1}{\varepsilon_2}\right), \tag{11.16}$$

and if the exposure time is the same for both regions, as is highly likely in practice, then

$$C_s = G \log\left(\frac{I_1}{I_2}\right), \tag{11.17}$$

provided that the operating point is on the linear portion of the curve of density against logarithm of exposure. When the radiation passes through inhomogeneous material, the values of I_1 and I_2 at the film will be different, and hence the film densities at these locations will be different.

11.3.3.2 Effects of Spatial Variation in Mass Attenuation Coefficients

If two regions of the material under test have different mass attenuation coefficients, μ_1/ρ_1 and μ_2/ρ_2, then the intensity of radiation passing through these regions will be

$$I_1 = I_o \exp\left(-\frac{\mu_1}{\rho_1}m_1\right) \quad \text{and} \quad I_2 = I_o \exp\left(-\frac{\mu_2}{\rho_2}m_2\right), \tag{11.18}$$

where m_1 and m_2 are the masses per unit area of the test material. Differences in I_1 and I_2 can be indicative of flaws in the material.

Because we have an equation for radiographic contrast in terms of the intensities of the incident radiation, we can write

$$C_s = G \log\left(\frac{\exp\left(-\frac{\mu_1}{\rho_1}m_1\right)}{\exp\left(-\frac{\mu_2}{\rho_2}m_2\right)}\right) = G\left(\frac{\mu_2}{\rho_2}m_2 - \frac{\mu_1}{\rho_1}m_1\right), \tag{11.19}$$

which means that to maximize contrast, we need to maximize the differences in the terms, $\frac{\mu_2}{\rho_2}m_2$ and $\frac{\mu_1}{\rho_1}m_1$. As μ_1 and μ_2 vary with energy of the incident x-rays, there will be an optimum x-ray energy or wavelength for obtaining maximum radiographic contrast.

11.3.3.3 Determination of Optimum Exposure Time for a Radiograph

Determination of optimum exposure for a radiograph is a complicated process [11]. In order to calculate the optimum exposure time for a radiograph, it is necessary to know the rating of the x-ray source in kV (or the photon energy of the x-rays) and the current supplied to the x ray set (which determines the intensity), the type of film (particularly the film gradient), the size of the x-ray source, the x-ray attenuation coefficient of the material, and the average thickness of the material.

11.3.3.4 Photon Energy

The x-ray photon energy can be calculated from the voltage rating of the x-ray set using the empirical equation:

$$\text{Peak energy} = \text{nominal voltage rating} \times 0.67. \tag{11.20}$$

11.3.3.5 Absorption or Attenuation Coefficient

The attenuation coefficient for the material of interest at the given x-ray photon energy is obtained from standard tables of these values. If necessary, the linear attenuation coefficient can be calculated from the mass attenuation coefficient using the equation:

$$\mu = \mu_m \rho. \tag{11.21}$$

11.3.3.6 X-Ray Intensity

The intensity of x-rays passing through the material (both with and without a flaw) is then calculated from the attenuation equation. The thicknesses used for this calculation will be the average thickness, x_{ave}, of the test material, and this thickness minus the size d of the flaw to be detected,

$$I_1 = I_o \exp(-\mu x_{ave}) \tag{11.22}$$

$$I_2 = I_o \exp(-\mu(x_{ave} - d)). \tag{11.23}$$

11.3.3.7 Radiographic Contrast

The radiographic contrast can be found from the values of intensities and the film gradient G:

$$C_s = G \log_{10}\left(\frac{I_1}{I_2}\right). \tag{11.24}$$

11.3.3.8 Film Unsharpness

The film unsharpness is found from the graph of U_f against x-ray energy as shown in Figure 11.3. Under optimum detection conditions, this should be equal to the geometrical unsharpness, U_g, calculated from the source size, the source-to-object distance, and the object-to-film distance,

$$U_g = \frac{F}{L_o} = U_f. \tag{11.25}$$

This allows the source-to-film distance, $L_o +$, to be optimized.

11.3.3.9 Equivalent Thickness of Steel

Exposure times are often tabulated or graphed in terms of data for steel, and so a thickness of steel that is equivalent to the test object needs to be found. For the given equivalent thickness of steel and the x-ray source rating in kV (or the x-ray photon energy in keV), an exposure in *milliamp minutes* is obtained from the exposure chart. If the current to the x-ray source is known, the required exposure time can be calculated.

11.3.3.10 Correction for Film Density

Adjustments are then made for the film density from the exposure chart. For example, if the film density is $D = 1, 2,$ or 3, the exposure time will be different, with higher density films needing lower exposure times.

11.3.3.11 Normalization for Source-to-Film Distance

The exposure charts are usually given for source-to-film distances of 1 m. If this distance is different, then a correction can be made for the different intensity, using the inverse square law. If the source-to-film distance is doubled, the exposure time needs to increase by a factor of 4:

$$I(L_o +) = I(1) \frac{1}{(L_o +)^2} \tag{11.26}$$

$$t(L_o +) = t(1) \cdot (L_o +)^2. \tag{11.27}$$

EXERCISES: RADIOGRAPHIC TESTING METHODS

11.1 A satisfactory film density had been obtained using contact radiography with a 20 min exposure time for a 25 mm thick specimen at a 700 mm source-to-film distance, with a 4 mm × 4 mm, 10 Curie iridium–192 gamma source. However, the iridium source is no longer available, and only a 0.5 mm × 0.5 mm, 1 Curie iridium–192 gamma source is available. If the same geometrical unsharpness is to be obtained, what exposure time will be needed?

11.2 From the radiographic film characteristics shown in Figure 11.10, determine:

1. The average gradient, G, between densities 1.5 and 2.5;
2. The exposure time needed to get a density of 2.5, if a 60 sec exposure gives a density of 1.3.

11.3 A contact radiograph was made of a pipe weld using a 2.2 Curie, cobalt–60 gamma source at the center of a 1.2 m outside diameter pipe. The exposure time was 2.5 h. A smaller pipe of the same wall-thickness and outside diameter 0.45 m is to be radiographed in the same way, using a similar source with smaller size. This source has a strength (known strictly as the *activity*) of 900 mCi. What will the new exposure time be? What is the largest source size that could ideally be

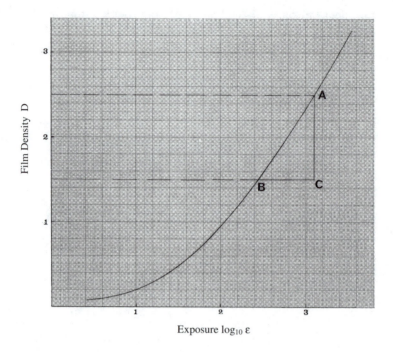

FIGURE 11.10 The dependence of radiographic film density on the logarithm of exposure.

used to radiograph the second pipe at this radial distance, if the pipe wall thickness is 18.75 mm and if the geometric unsharpness must not be more than 0.2 mm? Give the source size F in millimeters.

REFERENCES

1. Casanova, E.J., Some important basics in radiography, *Mater Eval* 44, 151, February 1986.
2. Munro, J.J., New and proposed standards for gamma radiography equipment, *Mater Eval* 39, 626, 1981.
3. Iddings, F.A., Radiation detection for radiography, *Mater Eval* 59, 926, August 2001.
4. Halmshaw, R., *Industrial Radiology: Theory and Practice*, Chapman and Hall, London 1995.
5. Ciorau, P., A contribution to the testing and classification of industrial radiographic films, *Nondestr Test Eval Int* 23, 152, 1990.
6. Notea, A., Evaluating radiographic systems using the resolving power function, *Nondestr Test Eval Int* 16, 263, 1983.
7. Aman, J.K., Fundamentals of radiography, Corney, G.M., McBride, D., and Turner, R.E., *Mater Eval* 36, 19, August 1978.

8. Graeme, W.A., Eizember, A.C., and Douglass, J., Digital image analysis of nondestructive testing radiographs, *Mater Eval* 48, 117, February 1990.
9. Stone, G.R., Gilblom, D., and Lehman, D., 100% x-ray weld inspection, *Mater Eval* 54, 132, 1996.
10. Warren Liao, T. and Gang Wang, Automatic identification of different types of welding defects in radiographic images, *Nondestr Test Eval Int* 35, 519, 2002.
11. McBride, D., Establishing standard x-ray exposures, *Mater Eval* 48, 1244, 1990.

FURTHER READING

Becker, G.L., *Radiographic NDT*, DuPont, Wilmington, 1990.
Bossi, R.H., Iddings, F.A., and Wheeler, G.C., *Handbook of Nondestructive Testing: Radiogrpahic Testing*, Vol. 4, ASNT, Columbus, 2002.
Halmshaw, R., *The Physics of Industrial Radiology*, Elsevier, New York, 1966.
Halmshaw R, *Capabilities and limitations of NDT. Part 3: Radiographic Methods*, Northampton; British Institute For Nondestructive Testing, 1989, 28 pp.
Halmshaw, R., *Industrial Radiology: Theory and Practice*, Chapman and Hall, London 1995.
http://www.ndt-ed.org/EducationResources/CommunityCollege/Radiography/cc_rad_index.htm.
Johns, H.R. and Cunningham, J.R., *The Physics of Radiology*, 4th ed., Thomas Publishers, Springfield, IL, 1983.

12 Thermal Testing Methods

Thermal testing is used mostly for detection of flaws in materials in which the local differences in thermal properties lead to measurable differences in temperature across a surface. Thermal inspection methods give rapid results and are often noncontact, although contact methods are also used. They are particularly useful for detection of subsurface flaws. Generally, these methods are better at detecting flaws in thin objects rather than thick objects, and are not so good at detecting deep-lying flaws. These methods can be applied to complex shapes and are mostly useful for single-sided inspections. Thermal inspection methods are particularly useful for composites or other structures made of dissimilar materials, where the variations in thermal properties are easily detectable.

12.1 HEAT TRANSFER

Thermal inspection consists of all methods which use heat-sensing devices in order to measure resulting temperatures and thermal gradients. This includes measurements of both transient thermal effects and steady state thermal conditions, either of which may be used to detect the presence of flaws, or thermal methods aimed at detecting unwanted distribution of heat flux from components. Thermography is the principal thermal testing technique in use at present. This provides a map of temperature variations across the surface of a test material [1], and can be used for monitoring temperature in steady state conditions to look for heat leaks, or for monitoring transients in temperature, which can be indicative of defects such as delaminations in the test specimen. In the case of thermal transients, the material is heated by subjecting it to a thermal pulse. Any flaws or inhomogeneities in the material lead to thermal gradients, which are different from those in the undamaged material.

12.1.1 THERMAL CONDUCTION

Heat will flow when a material is subjected to a temperature gradient. Thermal energy is transferred by three main mechanisms: conduction, convection, and radiation. In the case of materials evaluation, the main interest concerns the utilization of variations in thermal conduction. Under steady state conditions the heat flux (thermal energy per unit area per unit time) obeys Fick's first law, and depends on the temperature gradient and the thermal conductivity according to the equation,

$$\Phi = \kappa \frac{dT}{dx},$$ (12.1)

where ϕ is the heat flux, κ is the thermal conductivity, and dT/dx is the thermal gradient. The thermal conductivity of flaws in the material will be different from the parent material. Therefore, when subjected to a heat pulse, the transient temperature in the vicinity of flaws will be different from the parent material.

12.1.2 THERMAL RADIATION

One of the most convenient ways to heat a test material for nondestructive evaluation purposes is by exposure to infrared (thermal) radiation. This is radiation with wavelengths greater than 750 nm. The so-called near-infrared range has wavelengths up to 10 μm, while longer wavelengths are classified as the far-infrared range.

The intensity of infrared radiation from a material depends largely on two factors, the temperature of the material and its surface roughness. The wavelength of the infrared radiation from a material has a peak intensity that can be related directly to the temperature by an equation known as Wien's displacement law. In this case,

$$\lambda_{max} = \frac{b}{T},\qquad(12.2)$$

where λ_{max} is the wavelength of maximum emittance, T is the temperature in degrees Kelvin, and b is the Wien displacement coefficient, which is 2,897 μm $\cdot$ K^{-1}.

Practical applications of thermal methods for nondestructive evaluation require a surface with high emissivity to ensure that a large signal is obtained from the detector, and therefore a high signal-to-noise ratio is achieved. It is also helpful if the radiation levels from other sources are low. The reflection, absorption, and transmission of infrared radiation obey the same rules as for other forms of electromagnetic radiation. This means that good absorbers are good emitters of radiation, and good reflectors are poor emitters.

12.1.3 TEMPERATURE TRANSIENTS

If a test material is heated and then allowed to cool, it will obey a cooling law, in which the rate of change of temperature with time depends on the temperature difference between the material and its surroundings. This can be expressed in forms such as Newton's law of cooling, which states that the rate of change of temperature is proportional to the difference in temperature between the material, T, and the surrounding environment, T_s. Other variants of this law are similar, but the proportionality is not quite linear, so that the exponent in the equation is generally greater than one.

An expression of the generalized law of cooling is

$$\frac{dT}{dt} = C (T - T_s)^n,\qquad(12.3)$$

where the exponent n varies from 1 (Newton's law) to 5/4 (the so-called "five-fourths law").

12.1.4 HEAT FLOW

Heat flow is a diffusion process, and under steady state conditions it obeys Fick's first law, as described in Chapter 3. Under transient conditions, it obeys Fick's second law

$$\frac{\partial T}{\partial t} = -D\frac{\partial^2 T}{\partial x^2} , \tag{12.4}$$

where t is time and x is position.

If heat flows perpendicularly through a layer of material with thickness L and thermal diffusivity D, the transient temperature on the far side of the layer changes according to Fick's second law, from an initial temperature T_i to a final temperature T_f. The temperature $T(t)$ at any given time t is obtained by the solution of the above equation

$$\frac{T(t)-T_i}{T_f-T_i} = erf\left(\frac{L}{2\sqrt{Dt}}\right). \tag{12.5}$$

The time $t_{1/2}$ needed for the temperature of the far surface to change by half of this difference, $(T_f - T_i)/2$, is,

$$t_{1/2} = \frac{1.38\,L^2}{\pi^2 D} . \tag{12.6}$$

Typical values of the thermal diffusivity are $D = 100 \times 10^{-6}$ m^2 · sec^{-1} for metals, and $D = 1 \times 10^{-6}$ m^2 · sec^{-1} for composites.

12.1.5 THERMAL WAVES

Thermal wave interferometers comprise AC techniques which use phase sensitive detection with a lock in amplifier. A modulated laser beam is often used to heat the surface of the test material. Individual components can also be exposed to heat cycling. In these cases, the thermal waves can be described in a lossless situation by the following equation).

$$T(x, t) = T_o \exp(i(qx - \omega t)) , \tag{12.7}$$

but in a material there will be some decay of the thermal wave with distance, according to the more general equation for a damped wave

$$T(x, t) = T_o \exp(i(qx - \omega t)) \exp\left(-\mu x - \frac{t}{\tau}\right). \tag{12.8}$$

The terms μ and τ depend on the thermal conductivity of the material. However, flaws have a different value of thermal conductivity and, therefore, can be considered as regions with locally different values of μ and τ. As a result of these differences, the local values of $T(x, t)$ during transients will be different in the vicinity of flaws, compared with the parent material.

12.2 THERMAL INSPECTION PROCEDURES

There are two main classes of thermal inspection techniques: thermography, which maps temperature variations over a surface, and thermometry, which is aimed at actual measurement of temperature. The normal procedure is to heat the test material, but in some cases cooling is used. The heating process can utilize transients or steady state conditions. These normally involve (1) thermal pulses and (2) continuous thermal waves, respectively.

In the first case, the material is subjected to a short-duration pulse of heat, and the decay of the transients is used to evaluate defects. The transient thermal inspection methods rely on the time dependence of temperature or temperature gradients. The temperatures or temperature gradients are often determined by correlation with established calibration standards. The results are then interpreted to indicate the presence of flaws or other problem areas in the material. Laser pulse heating has been used for thermography of transients in temperature across a surface for crack detection [2].

In the second case, a steady-state condition is set up by repetitive or continuous long-term exposure to heat, which results in a steady-state condition that shows thermal features related to flaws. Direct detection of the thermal steady-state of components under their normal operating conditions can be an important consideration for operation of installations such as power plants [3], whether for heat leaks or for the detection of flaws.

12.2.1 CONTACT DETECTION METHODS

There are a variety of thermal measurement methods which depend on actual contact with the test material. These include measurements using thermocouples, thermistors, and thermographic methods which can include the use of coatings, paints, phosphors, and liquid crystals. Several different surface coating methods are available for detection of temperature gradients and temperature changes in materials. These include color change coatings, phosphor coatings, melting point coatings, and thermochromic liquid crystals (which give different colors across the whole visible spectrum at different temperatures).

12.2.2 NONCONTACT DETECTION METHODS

Noncontact thermal measurements are also widely used. These include thermometric techniques such as radiometers and pyrometers, and thermographic methods such as thermal cameras and infrared sensitive films. Detector arrays are often used to improve the signal levels, as the infrared intensity can be low for room temperature measurements.

12.2.3 THERMOMETRY

12.2.3.1 Contact Thermometry

These comprise traditional temperature measurement methods such as thermocouples, thermopiles, thermistors, and bolometers. A thermocouple is a pair of junctions between two metals, which give rise to a voltage that depends on the difference in temperature between the two junctions. A thermopile is a series of thermocouples that can thereby provide amplification of the signal in direct proportion to the number of thermocouple junctions. A thermistor is a temperature dependent resistor. A bolometer is a device that relies on the change in resistance of a sensor material (such as a fine wire or a thin film) with temperature.

Meltable substances, usually made of wax, with selected melting temperatures can also be used to determine temperature. These operate typically over the range 40 to 1700°C. In practice, the surface of the test material is marked with the wax (often these are supplied in the form of wax "crayons") prior to heating. When the test material reaches the preselected temperature, the wax melts. These methods are best adapted to determining when a particular preselected temperature is reached.

12.2.3.2 Noncontact Thermometry

These methods rely on the response of a sensor to infrared radiation. They are useful for remote sensing of surface temperatures and particularly for imaging.

Radiometers are thermal sensors which are mounted in a cavity such that the cavity focuses the heat radiation on the sensor. Radiometers without lenses respond to the total intensity of radiation incident on the surface of the sensor. In the case of radiometers with lenses, the range of wavelengths that they are sensitive to are restricted to the range of wavelengths that are transmitted through the lenses.

Pyrometers, also known as infrared thermometers, are used for higher temperatures, typically above 700°C. These are similar to infrared scanning devices, and rely on the measurement of temperature from the heat radiation emitted from hot components. The radiation is focused on to a sensor such as a thermocouple or thermopile, and the resulting voltage is measured.

12.2.4 THERMOGRAPHY

A thermograph is a map of temperature across a surface. Thermography can be used to detect variations in temperature across hot spots or just to locate them [4].

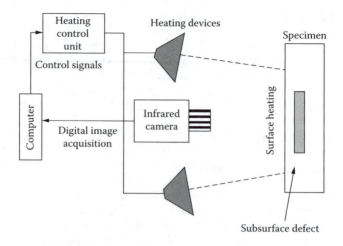

FIGURE 12.1 Setup for pulsed thermographic imaging system.

It can be used to detect flaws (transient thermography) [2], or even for stress detection [5].

The image can be produced by a variety of thermal imaging devices, depending in the temperature range of interest and the need for automated scanning, or by the use of hand-held devices [6]. The image is often produced using an infrared camera, as shown in Figure 12.1. The heat flux depends of course on the spatial homogeneity of the test specimen. The mechanical inhomogeneities lead to contrast in the resulting image. The instrumentation requirements include a radiative heating device under computer control, which gives a transient heat pulse.

The heat radiation emitted from the surface of the test material is detected by a camera, and the infrared image is digitally acquired and stored on a computer. The variation with time of the temperature at the surface of the material is then displayed to show regions of the material where the rate of heating or cooling is different from the rest of the material.

Thermography can be implemented in several different ways. These include pulsed thermography, modulated thermography (sometimes called lock in thermography), and pulse phase thermography [7]. Pulse thermography is the best known. This allows rapid noncontact evaluation of materials. The most sensitive technique is modulated thermography, which requires cyclic heating of the test material with detection based on variations in temperature occurring at the same frequency as the thermal excitation pulsing but with different phase lags, depending on the local variations in thermal properties. The signals can then be detected using a lock in amplifier (see Figure 12.2). This results in improved signal-to-noise ratio, because any signals that occur at other frequencies are rejected.

Interpretation of thermographic testing results can be enhanced through comparison of the results of practical inspections with modeling calculations. This can

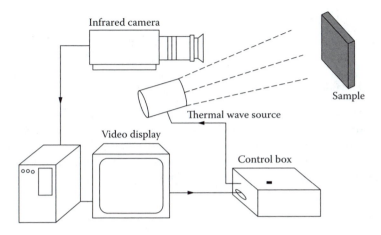

FIGURE 12.2 Setup for an oscillatory thermographic system that uses phase sensitive detection.

even include nonuniform heat distribution over the surface, as in the work of Maldague and Fortin [8], so that the presence of defects can be predicted from the thermal perturbations across a surface.

12.2.5 CONTACT THERMOGRAPHY

This involves coating of a surface with a temperature sensitive layer, such as a thermochromic film, thermal phosphors, heat-sensitive liquid crystals, heat-sensitive paint, or heat-sensitive paper.

12.2.5.1 Liquid Crystals

Cholesteric liquid crystals change color with temperature, and these can be used to indicate variations in temperature across a surface over a temperature range of −20 to 250 °C. Their response times are typically in the range of 100 msec, so that they can be used for dynamic imaging of thermal transients, as well as for static temperature imaging. The spatial resolution depends on the thickness of the layer, but a resolution of 0.02 mm can be achieved with a layer of the same thickness. For best performance, the test material should first be coated with a nonreflecting black paint before the temperature sensitive liquid crystal is applied.

12.2.5.2 Phosphors

Thermally quenched phosphors are materials that emit light in the visible range when exposed to ultraviolet light. These materials are useful over the temperature range 20 to 400 °C.

12.2.5.3 Heat Sensitive Paints

Heat sensitive paints can be used over the temperature range 40 to 1600 °C. They can give temperature determination with an accuracy of ± 5 °C.

12.2.5.4 Heat Sensitive Papers

Heat sensitive papers are also used for temperature detection. These papers are bonded onto the surface of the test material. They consist of a wax embedded in a paper substrate. The wax melts when a particular temperature has been reached. They are useful for indicating when a particular temperature has been reached, but are not useful for imaging of temperature variations.

12.2.6 NONCONTACT THERMOGRAPHY

In noncontact thermography, sensors are used to detect infrared radiation, which is then converted to a voltage signal and plotted as a function of position over a surface. Two main classes of sensors are used: photon effect devices, which depend on the wavelength of the radiation detected, and thermal effect devices, which depend on heating of the sensor. Far infrared imaging systems can have a temperature sensitivity of 0.1°C and a response time of 0.1 sec. Hand-held scanners can be useful for qualitatively detecting hot spots on components, but are not so good for quantitative measurements of temperature.

Noncontact thermography is used for such problems as detecting temperature variations across rolled steel strips. For this application, a thermographic camera is used. The actual detector or sensor is often a Cd-HgTe sensor with a bandwidth of 8 to 14 µm. Temperature resolution is typically 0.2°C, and the time constant is about 20 msec, so temperature transients can be detected by this method.

In practice, heat is supplied by a flash tube which can deliver high amounts of energy in very short pulses. Typically, 5 J.cm^{-2} pulse with a duration of a few msec (corresponding to a power density of the order of KW.cm^{-2}). This generates high initial temperatures on the exposed surface, which then decay with time.

The technique is good for detecting delaminations in materials and resin-rich regions in carbon fiber-reinforced polymer laminates. It is good for detecting flaws in composites such as honeycomb-bonded composites. One of the great advantages is that no contact is needed. One of the disadvantages of this technique is that it is not so good at detecting deeply buried flaws. Also, in some cases, the thermal shock arising from the sudden influx of heat energy can itself produce damage in the part, making the method no longer truly nondestructive.

Differential thermography can be used for detection of stress via the thermoelastic effect in which, under adiabatic conditions, changes in temperature of a region of material depend on the stress [5].

Thermography is perhaps the most widely used thermal testing method. Its applications range from small scale parts up to entire buildings, and the method can be entirely contact free, which brings advantages in many situations. Thermographic surveys of buildings can be appropriate both for identification of thermal leaks and for detection of structural problems [9].

12.2.7 APPLICATIONS OF THERMOGRAPHY

Mostly thermography is used for nonmetallic or poorly conducting materials, particularly composites and laminates [10]. The technique also finds widespread applications to plate glass, automobile windshields, nonmetallic coatings (uniformity tests), plastics and polymers, and in paper rolling and drying. Detection of flaws in concrete structures can also be achieved through the use of thermography [11, 12].

12.2.8 SONIC INFRARED INSPECTION

One of the recent thermal techniques to emerge is a method that uses high power ultrasonic excitation and infrared detection to locate defects [13]. The basic idea is that when a material is subjected to a short (~1 sec) intense burst of ultrasound (at 20 to 40 kHz), any cracks in the material will move, inducing frictional heating. This is a further development of an earlier method known as "vibro-thermography," based on the same principle. Consequently, local variations in temperature in the material, which can be located using an infrared camera, are indicative of the presence of cracks. Research has shown that the use of a range of frequencies of ultrasound is more effective in raising the temperature in the vicinity of cracks than is the use of a single ultrasound frequency [14]. This is because a spectrum of frequencies excites a larger fraction of the vibrational modes of the specimen, including those which are most efficient at causing differential motion across the crack. The method is proposed as an alternative to both fluorescent particle inspection and magnetic particle inspection.

REFERENCES

1. McQueen Smith, B., Condition monitoring by thermography, *NDT Int* 11, 121, 1978.
2. White, G. and Torrington, G., Crack detection and measurement using laser pulse heating and thermal microscopy, *Mater Eval* 53, 1332, December 1995.
3. Teich, A.C., Thermography for power generation and distribution, *Mater Eval* 50, 658, 1992.
4. Lucier, R.D., Recent work in infrared thermography, *Mater Eval* 49, 856, 1991.
5. Sandor, B.I., Lehr, D.T., and Schmid, K.C., Nondestructive testing using differential infrared thermography, *Mater Eval* 45, 392, 1987.
6. Newitt, J., Application specific thermal imaging, *Mater Eval* 45, 500, 1987.
7. Carlomagno, G.M. and Meola, C., Comparison between thermographic techniques for frescoes NDT, *Nondestr Test Eval Int* 35, 559, 2002.
8. Maldague, X. and Fortin, L., Aspects of thermal modeling for a thermographic NDT inspection station, in *International Advances in Nondestructive Testing*, McGonnagle, W., Ed., Gordon and Breach, New York, 1991.

9. Titman, D.J., Applications of thermography in non-destructive testing of structures, *Nondestr Test Eval Int* 34, 149, 2001.

10. Bouvier, C.G., Investigating variables in thermographic composite inspections, *Mater Eval* 53, 544, 1995.

11. Clark, M.R., McCann, D.M., and Forde, M.C., Application of infrared thermography to the non-destructive testing of concrete and masonry bridges, *Nondestr Test Eval Int* 36, 265, 2003.

12. Maierhofer, C., Brink, A., Pollig, M., and Wiggenhauser, H., Detection of shallow voids in concrete structures with impulse thermography and radar, *Nondestr Test Eval Int* 36, 257, 2003.

13. DiMambro, J., Ashbaugh, D.M., Han, X., Favro, L.D., Lu, J., Zeng, Z., Li, W., Newaz, G.M., and Thomas, R.L., The potential of sonic IR to inspect aircraft components traditionally inspected with fluorescent penetrant and or magnetic particle inspection, *Rev Prog Quant NDE*, 2003 or 2004.

14. Han, X., Islam, M.S., Favro, L.D., Newaz, G.M., and Thomas, R.L., Simulation of sonic IR imaging of cracks inmetals with finite element models", *Rev Prog Quant NDE*, 2006.

FURTHER READING

Davis, J.R., (Ed.), *Metals Handbook — Desk Edition*, 2nd Ed., ASM, Materials Park, OH, 1998.

Maldaque, X.P.V., *Handbook of Nondestructive Testing: Infrared and Thermal Testing*, Vol. 3, ASNT, Columbus, 2001.

Reynolds, W.N., *Capabilities and limitations of NDT. Part 8: Optical and Infrared Methods*, British Institute for Nondestructive Testing, Northampton, 1989.

Saintey, M.B., Evaluation of the Use of Thermal Techniques for the Nondestructive Testing of Sprayed Coatings and other Materials, Ph.D. thesis, University of Bath, 1995. 252 pp.

Thomas, R.L. et al., Thermal wave techniques, *Rev Prog Quant NDE,* Vol. 16, 353 et seq., 1997.

13 Destructive vs. Nondestructive Testing

Nondestructive testing, while of obvious benefit in the industrial environment, is not the only approach. Traditionally, most materials evaluation methods involve the use of test samples cut from the materials or components of interest. Alternatively, representative samples taken from a batch of nominally identical specimens may be subjected to laboratory testing. In some cases, destructive tests are more economical than nondestructive tests because the cost of the destructive test is usually lower. Therefore, for low-cost parts, the sacrifice of a few components can be tolerated. On the other hand, for high-added-value parts, the cost of sacrificing some parts may outweigh the additional cost of nondestructive testing, and even if it does not, the use of representative samples is founded on the uncertain assumption that the remaining samples will behave in the same way. Additionally, in some applications it is necessary to directly test the actual part that will be in service. This chapter also looks at materials characterization as a special case of testing, which, because of its different objective of determining materials properties instead of defect detection and failure, can be distinctly different from nondestructive evaluation.

13.1 TESTING OPTIONS

There are three components to nondestructive testing [1]: (1) indirect measurement of some materials property, (2) correlation of nondestructive tests with the property of interest, and (3) judgment concerning the serviceability of the part.

The correlation aspects of a test usually require the use of calibration standards which act as a reference [2]. Usually, the calibration specimens are simplified parts, (e.g., flat-bottomed holes) for ultrasonic calibrations. Although the use of these reference standards is never an exact representation of how the measurements should look on a real specimen, the measurements on reference standards provide a calibration of how the results should look under ideal conditions.

13.1.1 TESTING VS. NO TESTING

A very basic question to be asked in each case is whether to inspect or not. This can be best illustrated through an example.

WORKED EXAMPLE

The cost of inspecting a plant is $50,000/d, the time needed for inspection is 10 d, and the cost of downtime at the plant is $400,000/d; the cost of a failure at the plant has been calculated to be $500,000,000, and the cost of replacement parts is $20,000,000. If the probability of failure is 1% without inspection, but is effectively zero after inspection, is it economic to inspect the plant? Is it economic to inspect if the probability of failure is 0.8%?

Solution

Inspection cost = ($50,000 + 400,000) × 10 = $4,500,000. The cost of failure = $520,000,000. If the probability of failure is 1%, the probable cost of failure = cost of failure times probability of failure = $5,200,000. On this basis, economics dictates that it is cost-effective to inspect the plant. If the probability of failure is 0.8%, the probable cost of failure = $4,160,000. On this basis, economics suggests that it is not cost-effective to inspect the plant, unless a cheaper method of inspection can be found.

We see from the preceding paragraph that the decision can be affected by the probability of failure. As the probability of failure decreases, there is less incentive to inspect. Also, as the cost of failure increases, there is more incentive to inspect.

13.1.2 FACTORS TO CONSIDER IN SELECTING TESTS

Economics: Destruction of high-added-value parts may be unacceptable, which indicates that nondestructive testing is needed in these cases. It follows that more testing can be justified for high-value components if it reduces the probability of failure and thereby the probable cost of failure. Speed of inspection and cost of downtime in the cases where in-service inspection is not possible are also important economic considerations.

Safety: The cost of failure can become very high if there are safety issues or potential injuries or fatalities that would be involved if there were a failure. Safety critical parts are therefore like high-value parts because of the likely cost of failures, even if their initial production costs do not put them into this category. Safety issues can turn into economic issues as a result of fines, litigation, etc.

13.1.3 DESTRUCTIVE TESTING VS. NONDESTRUCTIVE TESTING

Nondestructive testing can be defined [3] as "the testing of a specimen that determines its serviceability without damage that could prevent its intended use." However, the procedures in nominally nondestructive tests must be followed carefully; otherwise, the tests can have deleterious effects on the parts being tested, or can invalidate the results of the tests [4].

Furthermore, nondestructive testing is not the only possibility when it comes to evaluation of materials. Destructive testing is an important alternative to nondestructive testing, and the class of destructive testing methods also has its place in the arsenal of techniques that can be brought to bear on a problem.

Destructive tests are usually viable in cases where the parts to be tested are cheap. This is often the case in mass-produced parts, where large numbers of nominally identical parts are produced. This allows batch testing of the parts in which a representative subset of the parts is collected for destructive testing. This type of testing is based on the assumption that all the other parts will behave in the same way as the ones being tested — in other words, the samples need to be truly representative of the larger group, so that the results can be carried over to the rest of the group.

Destructive tests include many of the traditional methods of assessing materials, including sectioning and polishing, together with microscopy, either optical or electron microscopy, sectioning for use with x-ray diffraction, and machining of samples for tensile, impact, fatigue, or compact tension tests. The following is a list of some standard destructive tests that are routinely used on materials:

- Tensile testing
- Fatigue testing
- Charpy impact testing
- Creep testing
- Metallography (sectioning, polishing, and viewing)
- Hardness testing
- Corrosion testing
- Compact tension testing (for crack growth)

Nondestructive testing is therefore mostly directed towards expensive, high-added-value parts, in which the cost of production is too great to allow some of the parts to be deliberately destroyed as part of the assessment. Nondestructive testing is much preferred on the production side of industry for online evaluation of parts coming from a manufacturing line, and is the only realistic option for in-service evaluation of parts in which they continue to perform their normal function during the inspection.

13.1.4 ECONOMICS OF TESTING

In the final analysis, nondestructive testing is an economic consideration [5]. The use of nondestructive testing to validate a part adds to its value by providing increased confidence in its ability to perform its intended function [6].

In the following sections, some simplified equations and examples are given as a guideline to economic considerations in nondestructive and destructive testing. These examples should not be taken too literally because, for example, it is very difficult to know in practice what the probability of failure will be, either before or after inspection, and therefore exact calculations are difficult to make. It is taken for granted that the inspection will have some beneficial effect on the probability of failure, either through rejection of bad components, or validation of good components for actual use.

13.1.4.1   Added Value vs. Cost of Nondestructive Testing

In quality control and quality assurance applications, the manufacturer may just need to know that the product meets certain specifications. No failure or safety issues may be involved. However, there has to be some added value associated with performing the test. If the initial value of the part is A\$, and the final value of the validated part is B\$, then the added value is B\$ $- A$\$. If the cost of conducting the nondestructive inspection for validating the part is C\$, then this inspection only makes economic sense if the added value exceeds the cost of the test that is to be used for the purpose of quality control/assurance,

$$C\$ < B\$ - A\$. \tag{13.1}$$

13.1.4.2   Added Value vs. Cost of Destructive Testing

In destructive tests, a fraction f of the parts that are tested is lost. So if a company performs such tests for quality assurance, then the above inequality needs to be modified to allow for the cost of the parts that are destroyed. For the tests to be economically justified, the following inequality must be satisfied,

$$C\$ < (1-f) \cdot B\$ - A\$. \tag{13.2}$$

For high-added-value parts, the fraction that are lost to destructive testing needs to be less than for low-added-value parts, because as f increases, then for large B\$, this rapidly decreases the total on the right-hand side of the inequality, making testing economically unviable sooner. In fact, even if the cost of the test is negligible, the destruction of a fraction f of the samples makes the testing uneconomic once the condition $(1 - f)B\$ < A\$$ is met.

13.1.4.3   Profit vs. Cost of Failure

In some quality assurance cases with mass-produced low-cost parts where there are no safety issues involved, it makes economic sense to conduct only minimal tests, or not to conduct any testing at all ($C\$ = 0$) and to just let the part fail. Even in this case, there are conditions under which such a strategy is economically justified, and this will be that the net profit B\$ $- A$\$ exceeds the probable cost of failure. If the cost of failure is D\$ and the probability of the part failing is p,

$$C\$ + pD\$ < B\$ - A\$. \tag{13.3}$$

13.1.4.4   In-Service Testing

In nondestructive testing for in-service components, there can be no discussion of initial cost and final cost of the components, but the discussion can be about pre- and post-tested components. If nondestructive testing can reduce the probability of failure from p_i to p_f, then for the test to be economically justified, the cost of the test must be less than the decrease in the probable cost of failure,

$$C\$ < (p_i - p_f)D\$. \tag{13.4}$$

13.1.4.5 Added Value vs. Costs of Nondestructive Testing and Failure

In the more general case, if nondestructive testing is performed at a cost of $C\$$ to reduce the probability of failure from p_i to p_f, then the inspection still only makes economic sense if the following inequality is satisfied:

$$C\$ < B\$ - A\$ + (p_i - p_f) \cdot D\$. \tag{13.5}$$

Clearly, the more effective the nondestructive test is in reducing the probability of failure p_f, the more expense that can be justified on the testing. Also note that this inequality shows that there are going to be conditions under which nondestructive testing is not economically justified.

13.1.4.6 Destructive Tests with Cost of Failure

The argument that destructive tests can reduce failures is dependent on the assumption that the parts tested and destroyed are identical to the parts that are not tested. In many instances this is questionable, particularly in the case of failures due to defects that may be present in only a fraction of the parts. In other cases this argument may be valid, for example, when the destructive tests are designed to establish intrinsic properties of the material, such as ultimate tensile stress.

If the associated cost of failure is $D\$$, and the probability of failure after the test is p_f, whereas the probability of failure before the test was p_i, then the following inequality needs to be satisfied for destructive tests in order that the testing is economically justified:

$$C\$ < (1 - f) \cdot B\$ - A\$ + (1 - f) \cdot (p_i - p_f) . D\$. \tag{13.6}$$

In the limiting case, as the added value $B\$ - A\$$ decreases towards zero, there is no incentive for testing.

WORKED EXAMPLE

Raw (untested) components are purchased at $500 each, whereas validated components can be resold at $600 each. If the cost of a failure is $2000, and the cost of a nondestructive inspection is $50, how low must the probability of failure of the validated components be to justify the inspection? What about the economic viability of selling of these parts without inspection or at zero cost of inspection?

Solution

The added value per part is $B\$ - A\$ = \$100$, whereas the cost of inspection accounts for $C\$ = \$50$ of this added value. The cost of a single failure is $D\$ = \$2000$, and the probable cost of failure is $pD\$$. Therefore, we need the following condition giving the upper limit to probability of failure that still makes the inspection economically viable

$$p < (B\$ - A\$ - C\$)/D\$ = \$50/\$2000 = 0.025$$
$$p < 0.025. \tag{13.7}$$

In the case of no inspection, this is equivalent to $C\$ = 0$, and therefore, the total profit on sale of a part is $\$100$. Therefore, this can tolerate a higher probability of failure and still be economically viable,

$$p < (B\$ - A\$ - C\$)/D\$ = \$100/\$2000 = 0.05. \tag{13.8}$$

So, above a 5% failure rate, the "no inspection" option is not economically viable.

13.1.5 ECONOMIC CONSIDERATIONS IN DESTRUCTIVE VS. NONDESTRUCTIVE TESTING

For either form of test we can define a net expected economic benefit (NEEB), which needs to be greater than zero for the test to be economically viable. For a nondestructive test, we can define the NEEB as

$$\text{NEEB} = B\$ - A\$ - C\$ + (p_i - p_f) \cdot D\$. \tag{13.9}$$

Whereas for a destructive test, we can define the NEEB of testing as

$$\text{NEEB} = (1 - f)B\$ - A\$ - C\$ + (1 - f) \cdot (p_i - p_f) \cdot D\$. \tag{13.10}$$

Generally, the value $C\$$ is higher for nondestructive testing than it is for destructive testing.

WORKED EXAMPLE

The cost of a power plant failure due to malfunction of a critical component is $\$1,500,000$. If by the use of nondestructive testing, which costs $\$100,000$, it can be established that a critical power plant component will not fail, whereas without testing the probability of failure over a given period is 10%, determine the NEEB of doing the nondestructive testing.

Solution

The net economic benefit is

$$NEEB = B\$ - A\$ - C\$ + (p_i - p_f) \cdot D\$. \tag{13.11}$$

In these cases, there is no change in value of the part, $B\$ - A\$ = 0$. So,

$$NEEB = -\$100,000 + (0.1) \cdot \$1,500,000 = \$50,000. \tag{13.12}$$

Therefore, it is economically justified to perform the tests.

13.2 MATERIALS CHARACTERIZATION

13.2.1 MATERIALS CHARACTERIZATION VS. NONDESTRUCTIVE EVALUATION

As measurement techniques, materials characterization and nondestructive testing have rather obvious areas of common interest, and so the boundaries between the two are somewhat blurred. Materials characterization refers to all of the measurement procedures that are conventionally used by material scientists and engineers to determine materials properties. This can include, for example, determination of mechanical properties, such as elastic modulus or tensile strength, electrical permittivity, magnetic permeability, thermal conductivity, acoustic velocity, microstructure, and so on.

13.2.2 INTRINSIC PROPERTIES VS. PERFORMANCE

The difference between material characterization and nondestructive testing, is that material characterization is aimed towards determination of materials properties, ideally under conditions in materials without problems such as flaws. Materials characterization is concerned with properties rather than performance. Materials characterization is not necessarily destructive, but neither is it restricted to being nondestructive and, in many instances, it can be destructive. For example, sectioning and polishing of materials for microstructural examination and preparation of tensile test samples for mechanical properties are standard destructive methods that are used in materials characterization.

Nondestructive evaluation and testing, on the other hand, is aimed towards finding the problems, such as flaws, cracks, and voids; or high dislocation densities and residual stress in materials, and attempting from this to determine the suitability of a particular component for performance in certain applications. In Figure 13.1, which has often been used to demonstrate the elements of materials science and engineering, it can be seen that the techniques that generally come under the heading of nondestructive evaluation and testing relate performance to

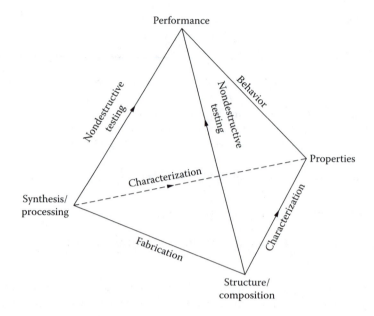

FIGURE 13.1 The four elements of materials science and engineering.

the three areas of synthesis/processing, structure/ composition, and properties, whereas material characterization links properties to the three other areas of synthesis and processing, structure/composition, and performance.

13.2.3 EXAMPLES OF CHARACTERIZATION METHODS

Almost any measurement technique that determines the properties of a material comes under the general category of material characterization. The broad classes are

- Mechanical tests
- Ultrasonic tests
- Optical tests
- Thermal tests
- Conductivity tests
- Magnetic tests
- Radiation tests

All the listed methods can be used for property evaluations, and in cases where the tests are both nondestructive and the properties can be linked to performance, they can also serve the purposes of nondestructive evaluation. In some cases, correlations between materials properties and failure-related mechanisms, such as the buildup of dislocations and residual stress, can be exploited to allow materials characterization of intrinsic properties to be used as an indirect measure of the condition of the material and its likely performance.

13.2.4 REMAINING LIFETIME AND FAILURE OF MATERIALS

Failure is one very critical aspect of performance. Failure of materials and the monitoring of approaches to failure are two of the main drivers behind the development of the field of nondestructive evaluation. However, failure itself is not the only issue, and may not be the main issue in particular instances where it may simply be sufficient to determine how a material performs during its lifetime under normal operating conditions.

The monitoring of materials properties can often be used to determine the condition of a material while it is operating under service conditions. In these cases, the correlation between properties that can be determined by standard materials characterization techniques (such as magnetic permeability, electrical conductivity, or ultrasonic velocity) and degradation of performance, may be all that is needed to determine that a component still has remaining service lifetime.

Ultimately however, it must be realized that the nondestructive evaluation of materials is also a statistical process, with probabilities of detection of defects less than unity [7], because of both the equipment and the operator. Therefore, there will always remain some uncertainty despite the increased levels of confidence provided by the test.

REFERENCES

1. Weinberg, M.H., Non destructive testing: what it is, what it involves, what it is used for and what it is not, *Mater Eval* 39, 111, 1981.
2. Cox, L.D. and Hammer, K.W., The NDT reference standard, *Mater Eval* 46, 702, 1988.
3. Iddings, F.A., NDT: what, where, why, and when, *Mater Eval* 56, 505, April 1998.
4. Laycock, S.A., What's destructive in nondestructive testing?, *Mater Eval* 41, 1108, 1983.
5. O'Connor, T.V. and France, C., Why do NDT?, *Mater Eval* 57, 705, July 1999.
6. Papadakis, E., Penny wise and pound foolish: the dangers of skimping on NDT, *Mater Eval* 60, 1292, 2002.
7. Cargill, J.S., How well does your NDT work?, *Mater Eval* 59, 833, 2001.

FURTHER READING

Johnson, R.V., The nondestructive testing workplace, *Mater Eval* 50, 926, 1992.
O'Brien, J. and Blum, W., Successful NDT procedure, *Mater Eval* 55, 1064, 1997.
Potter, R., Strategic planning for the NDT business, *Mater Eval* 61, 133, 2003.
Papadakis, E.P., Three philosophies: recalls, statistics and NDT, *Mater Eval* 61, 900, 2003.

14 Defect Detection

Defect detection is one of the main objectives of nondestructive evaluation. This chapter looks at the different types of problems that can arise as a result of inspection and how these are defined and classified. Signals from defects exhibit a statistical distribution and therefore, in practice, not all defects will be detected. Furthermore, inspections can also suggest the presence of flaws where there are none. This latter condition is known as a *false alarm*. This chapter discusses both the probability of detection and the probability of false alarms and how the choice of accept–reject criteria can affect these probabilities.

14.1 TERMINOLOGY FOR NONDESTRUCTIVE EVALUATION

14.1.1 DISCONTINUITY, IMPERFECTION, FLAW, AND DEFECT

In common parlance, these terms are used almost interchangeably as they relate to the condition of the material. However, precise definitions are needed for NDE [1,2].

A *discontinuity* is a lack of continuity of cohesion; an intentional or unintentional interruption in the physical structure of a material or component.

An *imperfection* is a departure of a quality characteristic from its intended condition.

A *flaw* is an imperfection or discontinuity that may be detectable by nondestructive testing and is not necessarily rejectable.

A *defect* is one or more flaws whose aggregate size, shape, orientation, location, or properties do not meet specified acceptance criteria and are rejectable.

14.1.2 NONCRITICAL FLAW

This relates to the condition of the material. A noncritical flaw is any discontinuity that does not adversely affect the performance or serviceability of a part.

14.1.3 CRITICAL FLAW

This relates to the condition of the material. A critical flaw is any discontinuity that adversely affects the performance or serviceability of a part.

14.1.4 INDICATION

An indication is the response or evidence from a nondestructive examination. It is an observation that could imply the presence of a discontinuity. The term relates to detected signals and is "measurement based," and so does not necessarily prove that a discontinuity is present. Any indication that is obtained requires interpretation and should be classified as relevant, nonrelevant, or false.

14.1.5 FALSE INDICATION

This is an indication that turns out not to be related to a discontinuity or imperfection. This could be the result of a spurious signal or the result of signal processing.

14.1.6 NONRELEVANT INDICATION

This is an indication that is caused by a condition or discontinuity that does not affect the performance or serviceability of the part. Therefore the part is not rejectable.

14.1.7 RELEVANT INDICATION

This is an indication that is caused by a condition or discontinuity that could affect the performance or serviceability of the part. Relevant indications require further evaluation.

14.1.8 INTERPRETATION AND EVALUATION

Interpretation is the process of determining whether an indication is false, nonrelevant, or relevant. Evaluation refers to the whole process of assessment of an indication to determine whether it corresponds to a flaw or not, whether this adversely affects performance or serviceability, and what action needs to be taken as a result of this.

14.2 PROBABILITY OF DETECTION

In this section, we examine the relationship between flaw size and signal levels. It is worth remembering that in NDE, decisions are made based on measurements of signals, rather than the actual flaw sizes, because flaw sizes are not measured directly, only inferred from indications obtained from inspections. No matter which inspection techniques are used, there will still be statistical limitations on the detection of defects caused by the fact that every measurement procedure has a distribution of signal levels to flaw size [3].

The probability of detection (POD) is usually determined from empirical studies on standard reference specimens. Analysis of the resulting data can proceed by different methods [4] but usually involves a statistical plot of signal amplitudes vs. flaw sizes.

The value of the POD depends on flaw size, and on the measurement technique and its ability to detect signals from flaws of different sizes. It says nothing about the actual size of a critical flaw. Therefore, if the critical flaw size is large, it may be easy to detect, conversely, if it is small, it may be difficult to detect. Basically, in situations where the critical flaws are larger, the indications from those flaws are also large enough to be easily seen. However, in situations where the critical flaws are smaller, and hence with smaller signals, there is a decreasing probability of detection. Under these latter conditions, some flaw signals may get overlooked because they are more difficult to detect.

Another approach to probability of detection is via model-based simulations, whereby predictions are made of expected signal levels for flaws of different sizes and orientations. The results of these simulations are then used in a similar way to the empirical data to show the variation of expected signal amplitude vs. flaw size.

14.2.1 Dependence of Signal Amplitude on Flaw Size

There is no completely general answer to the question of how the detected signal amplitude relates to flaw size. It is often assumed that the signal amplitude increases as the flaw size increases; and it is often implicitly assumed, for convenience of analysis, that the logarithm of signal amplitude increases linearly with the logarithm of flaw size [5]. In that case, the assumed relationship between flaw size and signal amplitude can be used to establish a threshold value of the detected signal that corresponds to a critical flaw size.

If we suppose that the relationship between flaw size, x, and flaw signal, V, is logarithmic within certain error limits, it is possible to write down an equation

$$\log_{10}(V) = \alpha \log_{10}(x) \pm \varepsilon, \qquad (14.1)$$

where α represents the basic relationship between signal voltage V and size of flaw x, whereas ε represents the uncertainty, because in practice there is bound to be a range of values of signal size for a given flaw size. This is represented in Figure 14.1.

In practice, the relationship between flaw size and signal amplitude does not often follow such a simple law, because, for example, even different orientations of a particular flaw can cause different signals to be detected in ultrasonic, eddy current, and radiographic inspection.

Another factor to consider is the signal-to-noise ratio (SNR). Noise has two origins: inherent noise on the detected signal, and noise arising from signal processing, such as analog-to-digital conversion [5].

14.2.2 Threshold Signal Level

At some point in the process of interpretation and evaluation, a decision has to be made about the amplitude of an indication above which the part is deemed flawed. This is not the same as deciding on the size and distribution of flaws,

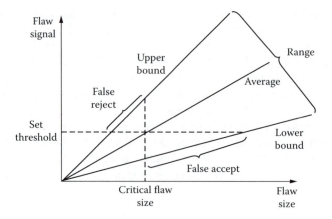

FIGURE 14.1 Dependence of signal amplitude on flaw size, assuming a basic logarithmic relationship with a certain range of signal amplitudes for a given flaw size.

but it is assumed that there is a relationship between flaw size and detected signal level.

A threshold signal level, V_{th}, is chosen so that any parts giving flaw signals above this level are rejected, and any that give signal levels below this are accepted. This is shown in Figure 14.1. It can be seen that this will not be able to completely separate the flawed and unflawed parts into just two correctly classified groups of rejects and accepts, but instead will result in additional groups of false accepts and false rejects.

WORKED EXAMPLE

For a certain type of flaw suppose $\alpha = 0.7$ when V is measured in volts and x is in meters and $\varepsilon = 10\%$. If we are looking for a critical flaw of size 1mm, what is the expected signal voltage and what is the range of possible voltages? What is the range of flaw sizes that will give the expected signal voltage?

$$\log_{10}(V) = 0.7 * \log_{10}(0.001) +/- 10\%$$
$$= -2.1 +/- 10\%. \qquad (14.2)$$

The expected signal voltage is

$$V_{ex} = 0.008 \text{ V.} \qquad (14.3)$$

The range is

$$\Delta(\log_{10}(V)) = -1.9 \text{ to } -2.3. \qquad (14.4)$$

With a band of uncertainty $\varepsilon = 10\%$, the range of signals for a 1 mm flaw will be from 0.0126 to 0.005 V.

The size d_{lo} of the smallest flaw that can give a signal as high as 0.008 V is

$$\log_{10}d_{lo} = (\log_{10}V_{ex} + 10\%)/\alpha$$
$$= -2.3/0.7 = -3.29 \tag{14.5}$$

$$d_{lo} = 0.00051 \text{ m.} \tag{14.6}$$

Therefore, all samples with flaws of size 0.51 mm and below will give signals of less than 8 mV and will not be detected with a set threshold of 8 mV. Therefore, all samples with flaw sizes below 0.51 mm will be correctly accepted.

The size d_{hi} of the largest flaw that can give a signal as low as 0.008 V is

$$\log_{10}d_{hi} = (\log_{10}V_{ex} - 10\%)/\alpha$$
$$= -1.9/0.7 = -2.71 \tag{14.7}$$

$$d_{lo} = 0.00194 \text{ m.} \tag{14.8}$$

Thus, all samples with flaws of size 1.94 mm and above will give signals of greater than 8 mV and will be detected with a set threshold of 8 mV. Therefore, all samples with flaw sizes above 1.94 mm will be correctly rejected.

For flaw sizes between 0.51 mm and 1.00 mm there will be, in addition to some correct accepts, also some false rejects caused by samples that give signal levels above the set threshold of 8 mV. Similarly, for flaw sizes between 1.00 and 1.94 mm there will be, in addition to some correct rejects, also some false accepts caused by samples that give signals below the set threshold of 8 mV.

14.2.3 PROBABILITY OF DETECTION: IDEAL CONDITIONS

The POD of a flaw is a vital piece of information for NDE. We may consider how the POD changes with the size of the flaw. It is reasonable to assume that as the flaw size increases, the POD of the flaw also increases toward the value 1. Under ideal circumstances there is: (1) no uncertainty about the relationship between flaw size and flaw signal, so the value of ε in the above equation would be zero; and (2) the variation of POD with flaw size is a step function in which the probability of detection suddenly increases from 0 to 1 at the flaw size x_{50}, as shown in Figure 14.2.

The value of x_{50} depends on the choice of the threshold V_{th}. There is one value of this threshold that can be chosen such that $x_{50} = x_c$, where x_c is the critical flaw size. Then the probability of a false reject is zero (because POD is zero) for flaws less than the critical size. Likewise, the probability of a correct reject is 100% (because POD = 1) for flaws greater than the critical size.

Under these idealized conditions, there is a 100% chance of getting relevant indications for flaws of size greater than the critical flaw size, and a 0% chance of getting relevant indications for flaws with size less than the critical flaw size. Therefore, there will be no false accepts and no false rejects. Also note that the

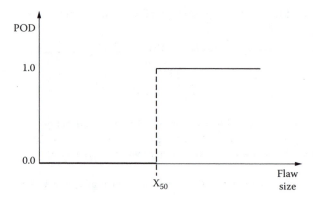

FIGURE 14.2 Variation of probability of detection, POD, with flaw size, x, under ideal conditions. If $x_c = x_{50}$ then POD = 0 for all $x < x_c$, and POD = 1 for $x > x_c$.

0% probability of detection for flaws with size less than x_c does not suggest a lack of sensitivity of the evaluation method for discontinuities below this size. Instead, the ability to detect discontinuities below x_c is almost always an advantage, but the graph actually suggests that although discontinuities below this size may be detectable, through interpretation of the signals it is possible to determine with complete accuracy whether the indications are relevant or nonrelevant. Such a situation almost never occurs in practice.

14.2.4 PROBABILITY OF DETECTION: REAL CONDITIONS

In reality, the probability of detection usually increases smoothly from 0% for small flaws up to 100% for large flaws [6], as shown in Figure 14.3. Schematically,

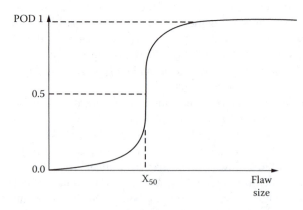

FIGURE 14.3 Schematic of the variation of the change of POD with flaw size.

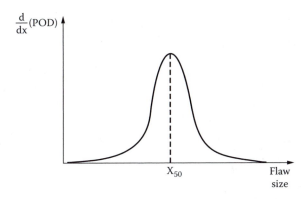

FIGURE 14.4 Schematic of the variation of the rate of change of POD, the derivative *d(POD)/dx*, with flaw size.

this shows the probability of detection increasing slowly with crack size for small cracks, and then there is a rapid increase centered on a flaw size x_{50}, which is the flaw size for which the POD is 0.5 (or 50%). At larger flaw sizes, the probability increases more slowly towards 1 with increasing flaw size. It should be recognized that defect detection is a statistical process, in which the probability of detection depends on two main factors: the limitations of the equipment and the limitations of the observer [7].

The dependence of POD on flaw size can be differentiated to give a distribution $d(POD)/dx$ vs. x which often looks like an error function or a normal (Gaussian) statistical distribution [3], as shown in Figure 14.4. In fact, although the POD vs. x curves usually have this type of appearance, and statistical tests can be performed to determine whether the data do follow a normal distribution, there is no fundamental justification for suggesting that it necessarily obeys a normal distribution. Other distributions such as lognormal, exponential or Weibull distributions also occur frequently, as discussed in Section 14.3.2. Figure 14.5 shows an actual distribution of the probability of detection for cracks in a metal component. Despite this, a normal distribution is often used to describe the variation of POD with flaw size.

There is no reason why the critical flaw size x_c should be the same as x_{50}, the flaw size at which the rate of change of POD is greatest. The flaw size which has a 50% probability of detection at a given threshold actually depends on the choice of the threshold voltage, and therefore x_c and x_{50} will coincide for one particular value of V_{th}.

Some generalized considerations for improving the accuracy of testing techniques have been proposed [8,9,10]. The proposed new testing methodology depends on some additional procedures to follow after all normal instrumentation optimizations have been completed. Part of this procedure consists of determining the response of the measuring equipment to calibrated "distorting excitations,"

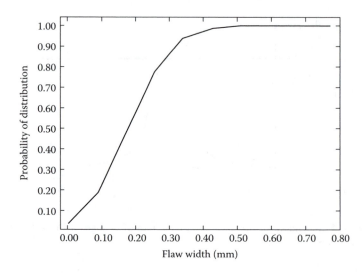

FIGURE 14.5 Actual probability of detection for cracks. For smaller cracks (<0.1 mm) POD is small, whereas for larger cracks (>0.4 mm) the POD approaches 1.

so that corrections can be made to measurement results. In addition, new procedures for data processing algorithms have been suggested and extrapolation methods proposed for improving interpretation of inspection data.

14.3 STATISTICAL VARIATION OF SIGNAL LEVELS

The statistical variation in signal levels from flaws of a given size means that there will always be some parts that are sorted into the wrong category, either as false accepts or false rejects. The probability of detection depends on the chosen threshold signal level. This can, of course, be set so that very few parts with critical flaws are accepted, but the consequences of such a decision are that there will also be a large number of parts without critical flaws which have signal levels above the set threshold, and are therefore classified incorrectly as rejects. This is shown schematically in Figure 14.6, where the distribution of signal levels for a given size of flaw includes a finite probability of false acceptance or false rejection.

14.3.1 MATHEMATICAL FORMALISM USING A NORMAL DISTRIBUTION

We now look at the mathematical description of probability of detection. For example, if the probability of detection POD increases as the flaw size x increases according to a normal or Gaussian distribution, it will have the form shown in Figure 14.3.

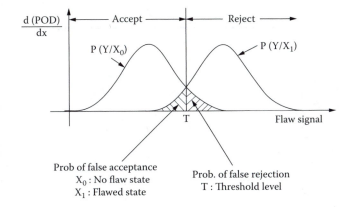

FIGURE 14.6 Dependence of the derivative of the probability of detection with flaw signal in two different cases. In one case the distribution of signals is for materials with no flaw, and the other case is the distribution of signals for materials with flaws. The selection of the threshold signal level for rejection V_{th} is such that, because of the statistical distribution of signals, even the unflawed materials will have a finite probability of giving a signal above V_{th}, therefore leading inevitably to some false rejects. Likewise, the flawed materials will have a finite probability of giving a signal below the threshold, so that there will be a finite probability of false accepts.

Putting this into a mathematical form, it will obey the following equation

$$POD(x) = \int_{-\infty}^{x} \frac{1}{\sqrt{2\pi}} \exp\left[-\frac{1}{2}\left(\frac{x - x_{50}}{\sigma} \right)^2 \right] dx, \qquad (14.9)$$

where x_{50} is the mean value, and σ is the standard deviation which controls the rate of change of probability as it passes through the mean value. This shows that for large flaw sizes, x, the probability of detection approaches 100%, whereas for small flaw sizes the probability of detection approaches 0. At $x = x_{50}$ the probability is exactly 50%.

The normal distribution curve, the familiar bell curve that is often associated with random distribution of quantities grouped around a mean, is the derivative of this function, and has the form shown in Figure 14.4.

$$\frac{d(POD)}{dx} = \frac{1}{\sqrt{2\pi}} \exp\left[-\frac{1}{2}\left(\frac{x - x_{50}}{\sigma} \right)^2 \right] \qquad (14.10)$$

The above equations give a mathematical model for quantitative analysis of probability of detection in terms of two degrees of freedom: the mean, x_{50}, and the width of the distribution, σ.

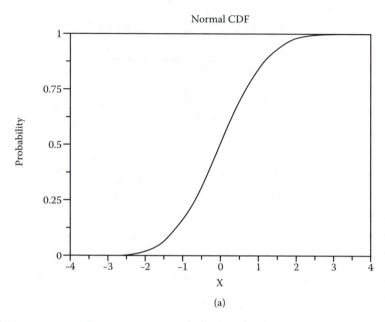

(a)

FIGURE 14.7 Plots of different statistical cumulative distributions $F(x)$ from the Engineering Statistics Handbook http://www.itl.nist.gov/div798/handbook/. These are a) Normal, b) Log normal, c) Exponential and d) Weibull distribution, with the following forms of equations:

a) $F(x) = \int_{-\infty}^{x} \frac{1}{\sqrt{2\pi}} \exp\left[-\frac{1}{2}\left(\frac{x}{\sigma}\right)^2 \right] dx$

b) $F(x) = \int_{-\infty}^{x} \frac{1}{\sqrt{2\pi}} \exp\left[-\frac{1}{2}\left(\frac{\ln x}{\sigma}\right)^2 \right] dx \quad x > 0, \sigma > 0$

c) $F(x) = 1 - \exp\left(-x/\beta \right) \quad x > 0, \beta > 0$

d) $F(x) = 1 - \exp\left(-x^r \right) \quad x > 0, \gamma > 0$

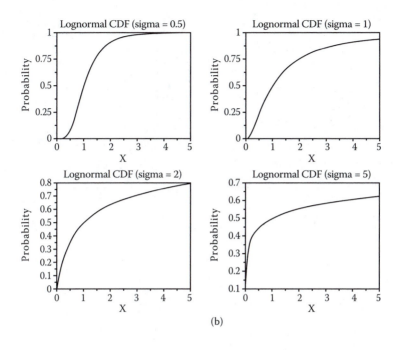

(b)

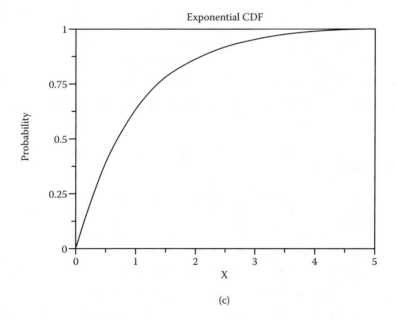

(c)

FIGURE 14.7 (Continued).

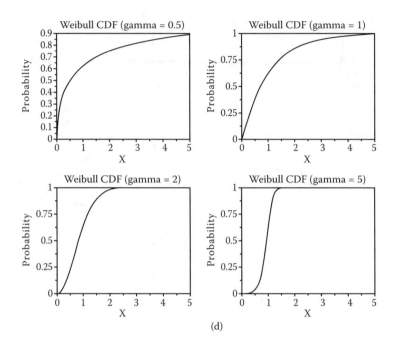

FIGURE 14.7 (Continued).

14.3.2 OTHER STATISTICAL DISTRIBUTIONS

Other distributions are possible and likely, depending of the specifics of the case under consideration [6]. Figure 14.7 shows the distribution curves of four of the most commonly occurring forms: normal, lognormal, exponential, and Weibull distributions.

REFERENCES

1. *Standard Terminology for Nondestructive Examinations*, E1316, Annual Book of ASTM Standards, Vol. 03.03 Nondestructive Testing, ASTM, West Conshohocken, Pennsylvania, 2001, pp. 645–683.
2. Thompson, R.B., Nondestructive evaluation and life assessment, *ASM Handbook: Failure Analysis and Prevention*, Vol. 11, ASM, Materials Park, Ohio, 2002, p. 269.
3. Papadakis, E.P., Three philosophies: recalls, statistics and NDT, *Mater Eval* 61, 900, 2003.
4. Berens, P.A., NDE reliability data analysis, *Nondestructive Evaluation and Quality Control: ASM Handbook*, Vol. 17, ASM International, Materials Park, OH 1989, p. 689.
5. Singh, G.P., Schmalzel, J.L., and Udpa, S.S., Fundamentals of data acquisition for nondestructive evaluation, *Mater Eval* 48, 1341, 1990.

6. Rummel, W.D., Probability of detection as a quantitative measure of non-destructive testing end-to-end process capabilities, *Mater Eval* 56, 29, January 1998.
7. Cargill, J.S., How well does your NDT work ?, *Mater Eval* 59, 833, 2001.
8. Matiss, I., New possibilities of increasing accuracy for nondestructive testing: Part 1, *Nondestr Test Eval Int* 32, 397, 1999.
9. Matiss, I. and Rotbakh, Y., New possibilities of increasing accuracy for nondestructive testing: Part 2, *Nondestr Test Eval Int* 32, 397, 1999.
10. Matiss, I. and Rotbakh, Y., New possibilities of increasing accuracy for nondestructive testing: Part 3, *Nondestr Test Eval Int* 32, 397, 1999.

FURTHER READING

Deng, Y., Liu, X., Zeng, Z., Udpa, L., Udpa, S., Shih, W., and Fitzpatrick, G., Numerical studies of magneto optic imaging for probability of detection calculations, *Proc Int Workshop on Electromagnetic NDE*, Vol. IX, IOS Press, June 2004, pp. 33–40.

Matzkanin, G.A. and Yolken, H.T., *Probability of Detection (POD) for Nondestructive Evaluation (NDE)*, Nondestructive Testing Information Analysis Center, Austin, TX, 2001.

15 Reliability and Lifetime Extension

This chapter discusses the decisions that may be made as a result of inspections for materials evaluation and nondestructive testing. These include the selection of a threshold level for acceptance/rejection of parts, the concept of false accepts and false rejects, and how these vary with the chosen threshold. The quantitative comparison of the effectiveness of different inspection procedures can be made using the relative operating characteristics curve. Finally, the idea of using materials evaluation or NDT for lifetime extension is discussed, where it is seen that the use of such inspections can lead to economic benefits, such as lower running costs due to less frequent replacement of parts, or timely replacement of parts, thereby avoiding costly failures.

15.1 RELIABILITY AND CRITERIA FOR DECISIONS

We now look at how the statistical distribution of signal levels for given flaw sizes leads to uncertainties in the interpretation of the condition of materials and, therefore, to situations where decisions are made that lead to false acceptance of flawed parts or false rejection of good parts. Both false accepts and false rejects are undesirable and costly. Reduction of the combined numbers of false accepts and false rejects is highly desirable, and this is achievable through the suitable choice of a threshold signal level.

In some cases it is necessary, for example, because of considerations of safety, to be conservative and just minimize the numbers of false accepts, while accepting the cost associated with increased numbers of false rejects.

15.1.1 Accept/Reject Criteria

This section deals with the signal levels and what actions need to be taken in terms of deciding on acceptable signal levels and the consequences of setting such signal levels higher or lower. It is important that we realize that in nondestructive evaluation we deal with detected signals, and we base decisions about serviceability of components on these detected signals and not directly on flaw size, because flaw size is only inferred from the measurements.

Accept/reject criteria are rules for establishing whether discontinuities can be tolerated without adversely affecting the performance or serviceability of a part. This usually means specifying how many defects are acceptable, and the combination of size and separation of these defects that can be tolerated in a component so that it is still able to perform its normal function in service.

Then indications or signal levels corresponding to these conditions are determined. It is also true that as the detected signal level increases, the probability that the part is still serviceable approaches zero.

> *False accept:* This means acceptance of a part on the basis of no indication, or an indication below a predetermined signal level, which ultimately is found to have a flaw present. This can be remedied by tightening the criteria for acceptance, which, in practice, will necessarily lead to more false rejects.
>
> *False reject:* This means rejection of a part on the basis of an indication above a predetermined signal level, which ultimately is found to have no flaw or a noncritical flaw. This can be remedied by relaxing the criteria for acceptance, which, in practice, will necessarily lead to more false accepts.

15.1.2 THRESHOLD SIGNAL LEVEL

In simple terms, the threshold signal level is set, as shown in Figure 15.1, so that above this level the part is rejected and below this level the part is accepted. More sophisticated schemes can be adopted in which there is an upper signal level beyond which all are rejected, a lower signal level below which all are accepted, and an intermediate range of signals for which further inspection is required. We will not discuss such procedures here.

It can be seen that, because of the range of signals between the lower and upper bounds it is possible that for a particular flaw size, there will be false accepts and false rejects based on the chosen accept/reject threshold.

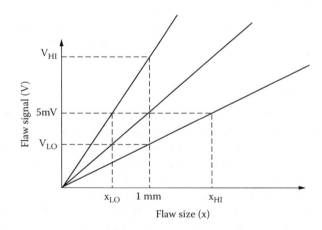

FIGURE 15.1 Linear variation of flaw signal amplitude with flaw size, which is often assumed for the purposes of analysis of defect detection, including setting of a threshold signal level for accept/reject decisions, as described by Equation 14.1. Threshold is set at 5 mV for a 1 mm flaw size.

TABLE 15.1
Decision Matrix

	Flawed	Unflawed	Sum
Indication/reject	Correct reject	False reject	Total rejects
No indication/accept	False accept	Correct accept	Total accepts
Sum	Total flawed	Total unflawed	Total

It can be seen from the simplified description of the relationship between flaw size and signal level in Figure 15.1, depending on the set detection threshold for signals and the expected upper and lower bounds for signals from a critical flaw, there will be circumstances under which specimens are rejected on the basis of a detected signal above the threshold, even though they actually meet the physical specifications for acceptance (false rejects); and specimens are accepted on the basis of a detected signal below the threshold, even though they do not meet the physical specifications for acceptance (false accepts).

15.1.3 DECISION MATRIX

The combination of conditions of the physical presence of flaw/no flaw in the material, together with the measurement parameter of indication/no indication, can be represented in a decision matrix as shown in Table 15.1. It can be seen from this matrix that there are some situations in which parts are accepted on the basis of no indication and no flaw present, or rejected on the basis of a positive indication and a flaw present. These result in correct decisions. Other situations, as shown in the matrix, can result in false accepts or false rejects, both of which are undesirable. The ideal is to minimize the numbers of parts that come into these two "false" categories.

WORKED EXAMPLE

In a batch of turbine blades, 50% contain critical flaws. If the percentage of false rejects was 10%, what was the percentage of false accepts? What is the percentage of correct accepts?

Solution

Here is the information that is available:

	Flawed	Unflawed	Sum
Indication/reject		10	
No indication/accept	?	?	
Sum	50	50	100

It follows from this that the percentage of correct accepts was 40%. However, there is not enough information available to give the percentage of false accepts.

WORKED EXAMPLE

In a batch of bolts, 75% give no indications of a flaw. If the percentage of correct accepts was 50%, what was the percentage of false accepts? In addition, if 15% of the bolts were correctly identified as flawed, what is the percentage of false rejects? From this information determine the percentage of bolts that were actually flawed?

Solution

	Flawed	Unflawed	Sum
Indication/reject	15	?	25
No indication/accept	?	50	75
Sum	?		100

From the above information, the percentage of false accepts was 25%, and the percentage of false rejects is 10%. The percentage of flawed bolts must have been 40%.

From the above examples it can be seen that for a series of flawed parts, the probability of false accepts (PFA) is related to the probability of detection (POD) by the equation PFA = 1 − POD. Likewise, for a series of unflawed samples, the probability of false rejects (PFR) is related to the probability of correct accepts (PCA) by the equation PFR = 1 − PCA.

15.1.4 FALSE ACCEPTS AND FALSE REJECTS

From the previous examples we deduce, however, that the PFA and PFR are not identical and are not even directly related. To know both, we need to have three pieces of information from among the following:

- Number of parts with flaws
- Number of parts with no flaws
- Number of parts giving indications
- Number of parts giving no indications
- Number of correct accepts
- Number of correct rejects

WORKED EXAMPLE

The POD of a certain class of cracks under a rivet head is 75%. A hundred parts are tested; of these, 80 do not contain critical defects. How many false accepts will there be?

Solution

Using the decision matrix, there are 20 with defects, of which 75% (a total of 15) will be detected, leaving 5 that are flawed but will not be detected.

	Flawed	Unflawed	Total
Rejected	15	?	?
Accepted	5	?	?
Total	20	80	100

Notice that insufficient information is available to fill in all the entries in the matrix, but such additional information is not needed to answer the question posed. Also notice that there is no way of telling from the information available what happens to the 80 parts without defects. Some of these will likely fall into the category of false rejects, but there is no relationship between the number of false accepts and the number of false rejects.

WORKED EXAMPLE

In the previous example, more information became available and it was found that 75 parts were found to be acceptable and 25 were rejected. How many false rejects were there? How many correct accepts were there?

Solution

The additional information can be added into the decision matrix

	Flawed	Unflawed	Total
Rejected	15	?	25
Accepted	5	?	75
Total	20	80	100

From this, it is very easy to see that the number of unflawed components that were incorrectly rejected was 10, and the number of correct accepts was 70.

15.1.5 DEPENDENCE OF PROBABILITY OF FALSE ACCEPTS ON THRESHOLD SIGNAL LEVEL

The PFA is determined by a combination of the chosen threshold signal level and the probability of detection of flaws. Once a signal threshold level is decided, all components with signal levels below the threshold are accepted, and all components with signal levels above the threshold are rejected. Therefore, for any given size of critical flaw, the higher the chosen threshold signal level, the more parts with critical flaws will be accepted. The variation of PFA with threshold signal level is shown in Figure 15.2.

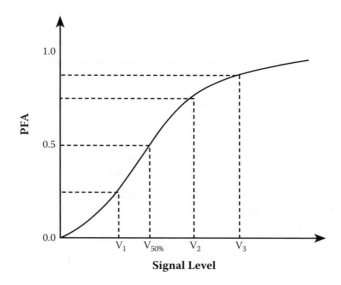

FIGURE 15.2 Variation of the probability of false acceptance (PFA) with signal size, which shows that if the accept/reject threshold is set too high, many bad components will be falsely accepted.

Assuming, as before, that the detected signal size V is proportional to the size of discontinuity x, we know that high detected signal levels are generally indicative of large numbers and sizes of flaws, and low detected signal levels are generally indicative of small numbers and sizes of flaws. For accept/reject decisions based on inspection results, we set a threshold voltage signal V_{th}, and then compare the detected signal with the threshold — primarily addressing the question of whether V is greater than or less than V_{th}. We can ask what happens if V_{th} is set at different levels. As the threshold signal level is decreased, the probability that the part will be accepted decreases towards 0%. On the other hand, as the threshold signal level is increased, the probability that the part will be accepted approaches 100%.

If $V_{th} = V_1$ and PFA $= 0.25$ and if this value of V_{th} corresponds to the critical flaw size, then 25% of the time there will be false accepts, meaning that out of every hundred parts which are sufficiently flawed that they do not meet specifications, 25 will nevertheless be accepted. If $V_{th} = V_2$, and PFA $= 0.75$ and if this value of V_{th} corresponds to the critical flaw size, then 75% of the time there will be false accepts. Taking a more extreme case of $V_{th} = V_3$, and PFA $= 0.9$ and if this value of V_{th} corresponds to the critical flaw size, then 90% of the time there will be false accepts.

It is clear that the number of false accepts increases as the threshold signal level V_{th} is increased, because more flaws are being ignored, and the parts containing these flaws are being accepted. To minimize the number of false accepts, the threshold signal level V_{th} has to be set as low as possible, but this will also have the unwanted effect of increasing the number of false rejects.

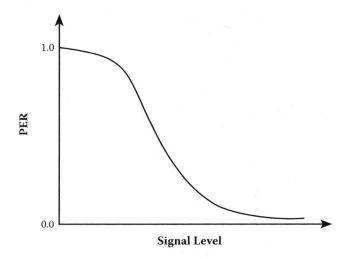

FIGURE 15.3 Variation of the probability of false rejection (PFR) with signal size, which shows that if the accept/reject threshold is set too low, many good components will be falsely rejected.

15.1.6 DEPENDENCE OF PROBABILITY OF FALSE REJECTS ON THRESHOLD SIGNAL LEVEL

A related viewpoint is to consider the probability of false rejects. There is no simple relationship just between the PFA and the PFR, because to calculate the latter from the former, we would need to know either the probability of correct accepts or the probability of correct rejects.

If a given signal size is observed to be at or above the threshold signal level, we must ask what is the probability that the indication is noncritical and that the part is, in fact, still serviceable. As the threshold signal level is decreased, the probability that the part will be rejected increases towards 100%. On the other hand, as the threshold signal level is increased, the probability that the part will be rejected approaches 0%. Figure 15.3 shows the variation of the PFR with signal size.

It is clear that the number of false rejects increases as the threshold signal level V_{th} is decreased, because fewer flaws are being ignored, and the parts containing these flaws are being rejected. To minimize the number of false rejects, the threshold signal level V_{th} has to be set as high as possible, but this will also have the unwanted effect of increasing the number of false accepts.

15.1.7 RELATIVE OPERATING CHARACTERISTICS: THE ROC CURVE

Ideally, one would like to have a high probability of correct rejects (PCR, which of course, is POD) and a low PFR. The plot of these quantities against each other is known as the relative operating characteristic or ROC. This is shown in Figure 15.4.

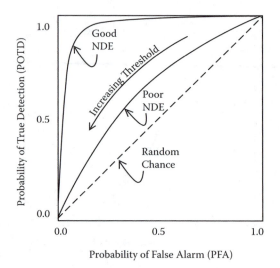

FIGURE 15.4 Relative operating characteristic (ROC) curve.

The POD depends on the chosen threshold for rejection. For a high threshold, there is a low POD but also a low PFR. This condition is represented by the curve in the bottom left-hand corner of the plot. For a low threshold, there is a high POD but also a high PFR. This condition is represented by the curve in the upper right-hand corner of the plot. These extremes will be true for any NDE technique and do not represent particularly useful operating conditions. The test of the effectiveness of an NDE technique is given by how it performs at intermediate thresholds. If it can give high probabilities of detection with low probabilities of false rejections, then the curve will tend towards the top left-hand corner of the plot, and operating under these conditions provides the best NDE capability.

When the POD equals the PFR, there is no added value arising from the nondestructive inspection. The more effective the nondestructive inspection, the further the ROC curve will be towards the top left-hand corner of the graph [1].

15.2 LIFETIME EXTENSION

This book began by introducing the concept of materials lifetime. We now look finally at the role of nondestructive testing in relation to this concept. Nondestructive testing and evaluation can be used to extend the safe operating lifetime of components by allowing replacement for cause, rather than replacement on a predefined schedule that is not related to materials condition and the state of degradation [2].

In practice, an economic case needs to be made for nondestructive testing and evaluation. This could be in terms of the added value of a component once it has been shown to be suitable for a particular application [3]. This may be achieved through materials characterization by showing that the component has properties that allow it to perform a particular function. It may be achieved by demonstrating that the material does not have flaws that will impair its function. Another way that this may be achieved is to show that a component that has been in service, and therefore has been subject to service aging, is still capable of continuing without suffering an impending failure. This last application can be broadened into the concept of *plant life extension*, in which the continued safe operation of an entire installation or plant is dependent on the continued safe operation of its various constituent components.

In plant life extension, as in lifetime extension of individual components, the questions that need to be asked are:

- How costly is a failure?
- How likely is a failure without nondestruction testing and evaluation?
- Can the nondestructive testing and evaluation detect the likely types of problems encountered?
- How costly is the inspection?
- Can nondestructive testing and evaluation lead to a reduction in the likelihood of failure, either through assurance of the integrity of the component, or by indicating the need for replacement?
- If the result of nondestructive testing and evaluation recommends replacement, how costly is the replacement?
- How costly is the downtime needed for inspection?

All of these factors need careful consideration and, in addition, there are broader strategic and socioeconomic considerations that apply [4].

15.2.1 Economic Considerations

Ultimately nondestructive testing and evaluation is an economic consideration [3]. As shown previously, there are inequalities that need to be satisfied to make the economic case. In plant life extension if the cost of failure is $D\$$, the probability of failure over a given time period is p, the cost of the inspection is $C\$$, and the cost of replacement parts is $E\$$, then we can calculate a probable cost associated with this which is $C\$ + pD\$ + E\$$.

An important question is what happens if inspections are not carried out ($C\$ = 0$). In this case, the frequency of replacement of the parts will have to be increased, which means an increased cost $E\$$. Alternatively, if there are no safety issues associated with failure, it may be possible to try to go without inspection or replacement, but because of the materials lifetime concept that we have discussed earlier, this will eventually lead to failure of the part. Allowing the part to fail means absorbing the cost of failure, $pD\$$, in which p without inspection is greater than p with inspection.

The economic role of nondestructive testing and evaluation should therefore be to minimize the sum of all the expenditures, $E\$ + C\$ + pD\$$, associated with operating a plant.

15.2.2 Retirement for Cause

An obvious economic benefit of nondestructive testing and evaluation is that it can be used to determine when a part needs to be replaced because of degradation, rather than just by setting up a predetermined retirement schedule for components [5]. Predetermined retirement schedules need to err on the side of safety, and this inevitably means that some parts will be removed from service before their actual service lifetime has been expended. Integrating this over a large number of low-cost parts, or even for just a few high-cost parts, there is an unnecessary cost resulting from too frequent replacement of parts.

Retirement for cause is a concept that has grown from a specific research program in which the replacement of parts for reasons arising from nondestructive inspection is seen to be more cost-effective, because parts are only removed from service when there are identifiable problems. In some cases, parts can continue to operate safely beyond their design lifetime, provided that it can be demonstrated they have not deteriorated beyond predetermined levels.

The use of periodic nondestructive inspections as part of a program of retirement for cause can be used to evaluate constituent components of larger structures to extend their lifetimes.

REFERENCES

1. Burkel, R.H., Choi, C.P., Keyes, T.K., Meeker, W.Q., Rose, J.H., Stuges, D.J., Thompson, R.B., and Tucker, W.T., A methodology for the assessment of the capability of inspection systems for detection of subsurface flaws in aircraft turbine engine components, U.S. Department of Transportation, Federal Aviation Administration Report No. DOT/FAA/AR-01/96, September 2002.
2. Engblom, M., Considerations for automated ultrasonic inspection of in-service pressure vessels in the petroleum industry, *Mater Eval* 47, 1332, 1989.
3. Papadakis, E.P., Penny wise and pound foolish: the dangers of skimping on NDT, *Mater Eval* 60, 1292, 2002.
4. Potter, R., Strategic planning for the NDT business, *Mater Eval* 61, 133, 2003.
5. Iddings, F., NDT: What, where, why and when, *Mater Eval* 56, 505, 1989.

Appendix: Solutions to Exercises

MECHANICAL PROPERTIES OF MATERIALS

2.1

1. From the stress–strain curve of 1095 steel, the elastic modulus Y is given by

$$Y = \frac{\sigma}{\varepsilon} = \frac{390 \times 10^6}{2 \times 10^{-3}} = 1.95 \times 10^{11}.$$

2. Draw a line parallel to the initial elastic region of the stress–strain curve cutting the x-axis at 0.002 strain. This line cuts the stress–strain curve at

$$\sigma_y = 550 \ MPa.$$

3. The nominal ultimate tensile strength is the maximum value of stress along the stress–strain curve

$$\sigma_y = 975 \ MPa.$$

Correcting this for the change in cross-sectional area gives

$$\sigma_y = \left(\frac{9.07}{8.18}\right)^2 \times 975 \ MPa = 1{,}119 \ MPa.$$

4. Ductility: The elongation of the material when it fails gives a breaking strain of $\varepsilon = 0.11$.
5. The percentage reduction in area =

$$\frac{final \ area - original \ area}{original \ area} \times 100\% = \frac{(8.18)^2 - (9.07)^2}{(9.07)^2} \times 100\% = -18.66\%.$$

2.2

In this case we need to start from the three-dimensional version of Hooke's law

$$e_x = \frac{1}{Y}(\sigma_x - v\sigma_y - v\sigma_z)$$

$$e_y = \frac{1}{Y}(\sigma_y - v\sigma_z - v\sigma_x)$$

$$e_z = \frac{1}{Y}(\sigma_z - v\sigma_x - v\sigma_y).$$

In the situation described, the constraint along the y-direction ensures that

$$e_y = 0,$$

whereas the lack of constraint along the x-direction ensures that

$$\sigma_x = 0.$$

Because we know that $Y = 200$ GPa, $v = 0.3$, and $\sigma_z = 200$ MPa, we can use the above equation for e_y to obtain

$$e_y = 0 = \frac{1}{200 \times 10^9}(\sigma_y - 0.3 \times 200 \times 10^6)$$

$$\sigma_y = 60 \, MPa$$

Likewise, for the strain in the z-direction, we can use the above equation for e_z

$$e_z = \frac{1}{200 \times 10^9}(200 \times 10^6 - 0.3 \times 60 \times 10^6) = 0.91 \times 10^{-3}.$$

The net result of the constraint along the y-direction is to reduce the strain along the z-direction.

2.3

Although no completely exact equation exists to relate ultimate tensile strength to hardness, some empirical relationships exist for particular materials, and in this case, Figure 2.12 can be used to determine hardness from tensile strength. For steels, a tensile strength of 1500 MPa corresponds to a Rockwell hardness, H_{RC}, of 45 and a Brinell hardness of BHN = 425.

From the equation for Brinell hardness,

$$BHN = \frac{P}{\pi D t} = \frac{2P}{\pi D\left[D - \sqrt{D^2 - d^2}\right]}.$$

Rearranging gives

$$d = \sqrt{\left[D^2 - \left(D - \frac{2P}{\pi D \cdot BHN} \right)^2 \right]}$$

$$d = \sqrt{\left[100 - \left(10 - \frac{2000}{4250\pi} \right)^2 \right]} = 1.72 \ mm.$$

Similarly,

$$t = \frac{P}{\pi D \cdot BHN}$$

$$t = \frac{1000}{4250\pi} = 0.075 mm.$$

SOUND WAVES: ACOUSTIC AND ULTRASONIC PROPERTIES OF MATERIALS

3.1

This is simply a problem in trigonometry. Whether there are reflections or not from the back wall does not matter for the standoff distance, which is the component of distance of the flaw parallel to the surface.

Let v be the ultrasonic velocity, s be the distance of the flaw along the path of the ultrasonic beam, d be the depth of the flaw, ℓ the standoff distance, and θ the probe angle.

$$2s = Vt$$
$$= 3000 \times 46.6 \times 10^{-6}$$
$$s = 70 \ mm$$
$$d = s \cos\theta$$
$$= 70 \times \cos 50$$
$$= 45 \ mm$$
$$\ell = s \sin\theta$$
$$= 70 \sin 50$$
$$= 53.6 \ mm.$$

3.2

The relationship between wavelength, velocity, and frequency is

$$V = \lambda \nu.$$

For longitudinal waves in steel:
If $\nu = 5 \times 10^6$ Hz and, in steel, v = 5,960 msec^{-1}, the wavelength is

$$\lambda = (5,960)/(5 \times 10^6) = 1.19 \text{ mm}.$$

For longitudinal waves in lead:
In lead, the velocity is v = 1960 msec^{-1} and, therefore,

$$\lambda = (1,960)/(5 \times 10^6) = 0.39 \text{ mm}.$$

For shear waves, these values of wavelength will be approximately half those of the longitudinal waves.
Steel:

$$\lambda = 0.6 \text{ mm}.$$

Lead:

$$\lambda = 0.2 \text{ mm}.$$

3.3

The wave velocity V is given by

$$V = \sqrt{\frac{Y}{\rho}} = 5000 \ m.s^{-1}.$$

The wavelength λ is given by

$$\lambda = \frac{V}{\nu} = 0.5 \ mm,$$

where ν is the frequency.
For the back wall reflection, the thickness d of the part is given by

$$2d = Vt$$

$$= 0.02 \text{ m}.$$

The thickness is therefore $d = 10$ mm.

Assuming no divergence of the wave front, the attenuation coefficient, μ, is

$$\mu = -\frac{1}{2d} \log_e \left(\frac{I}{I_0} \right) = 11 \, m^{-1}.$$

THERMAL PROPERTIES OF MATERIALS

4.1

Stress σ can be calculated from $\sigma = eY$, where e is the strain and Y is the Young's modulus.

The net stress will be

$$\sigma_{net} = \sigma_{Al} - \sigma_{Fe}$$

$$= e_{Al} Y_{Al} - e_{Fe} Y_{Fe},$$

and the strains can be calculated from

$$e = \alpha \, \Delta T,$$

where α is the thermal expansion coefficient. Therefore,

$$\sigma_{net} = (\alpha_{Al} Y_{Al} - \alpha_{Fe} Y_{Fe}) \, \Delta T$$

$$= ((26 \times 10^{-6})(0.7 \times 10^{11}) - ((12 \times 10^{-6})(2.0 \times 10^{11}))) 45$$

$$= 26.1 \text{ MPa}.$$

4.2

The heat flow through the wall of the pipe and through the insulation must be the same under steady-state conditions,

$$\frac{Q}{At} = \frac{K_{Fe}(T - \theta)}{x_{Fe}} = \frac{K_{ins}(\theta - 40)}{x_{ins}},$$

where K_{Fe} and K_{ins} are the thermal conductivities of the steel pipe and the insulation, x_{Fe} and x_{ins} are the thicknesses of the steel pipe and the insulation, T is the temperature inside the pipe, and θ is the temperature at the interface of the pipe and insulation.

We know that

$$\frac{Q}{At} = 15,000 \ W \cdot m^{-2}.$$

Therefore, with two simultaneous equations in two unknowns, T and θ, we can solve for both.

For the heat flow through the insulation,

$$\frac{4(\theta - 40)}{0.02} = 15,000,$$

and therefore $\theta = 115°C$.

For the heat flow through the steel pipe,

$$\frac{50(T - 115)}{0.008} = 15,000,$$

and therefore $T = 117.4°C$

4.3

First convert calories to joules. 5000 cal = 21,000 J. The change in temperature is given by

$$\Delta T = \frac{1}{m}\frac{\Delta Q}{C} = \frac{1}{2}\frac{-21,000}{500} = -21 \deg.$$

This puts the final temperature at −1°C, which is below the ductile-to-brittle transition temperature. So the material will be brittle, which could be a concern for continued operation.

ELECTRICAL AND MAGNETIC PROPERTIES OF MATERIALS

5.1

The frequency needed to give a depth of penetration of 3 mm can be obtained from the skin depth equation

$$\delta = \sqrt{\left(\frac{1}{\pi \nu \sigma \mu}\right)},$$

so that

$$\nu = \frac{1}{\pi \mu \sigma \delta^2} = \frac{1}{(3.142)(3.5 \times 10^{-4})(5 \times 10^5)(0.003)^2} = 202 \ Hz.$$

5.2

The impedance is given by

$$Z = \sqrt{\left[R^2 + (X_L - X_C)^2 \right]},$$

with
  $R = 10 \ \Omega$
  $X_L = \omega L = 0.314$
  $X_C = 1/(\omega C)$
  and $\omega = 2\pi v = 2,513$ rad/sec.

Therefore, to make the impedance 12 Ω, the capacitance needs to be

$$C = \frac{1}{2513} \times \frac{1}{0.31 + \sqrt{12^2 - 10^2}}$$

$$C = 57 \ \mu\text{F}.$$

At this frequency, the penetration depth is

$$\delta = \sqrt{\left(\frac{1}{\pi v \sigma \mu} \right)} = \sqrt{\frac{1}{(3.142)(400)(12.56 \times 10^{-7})(35 \times 10^6)}}$$

$$\delta = 0.00425 \ \text{m}.$$

5.3

The flux density, B, caused by a magnetic field, H, is given by

$$B = \mu_o \mu_r H,$$

and the field generated by a long solenoid is

$$H = ni,$$

so the required current is

$$i = \frac{B}{n \mu_o \mu_r} = \frac{1.1}{10,000 \times 12.56 \times 10^{-7} \times \mu_r}.$$

For the material with $\mu_r = 20$, this gives $i = 4.38$ A.
For the material with $\mu_r = 75$, this gives $i = 1.17$ A.

Assuming that the same field strength, H, is needed at the surface of the parts, then for a linear current passing down the center of the rod with radius a

$$i = \frac{2\pi a B}{\mu_o \mu_r}.$$

The material with $\mu_r = 20$ needs $i = 1376$ A.
The material with $\mu_r = 75$ needs $i = 367$ A.

EFFECTS OF RADIATION ON MATERIALS

6.1

The count rate is proportional to intensity, and this is governed by the equation

$$I(x) = I(0) \exp(-\mu x).$$

Through the full thickness of the lead container, $x = 0.050$m, whereas through the region with a 3-mm cavity, $x = 0.047$m. The fractional change in count rate is

$$\frac{\Delta N}{N} = \frac{\exp(-0.047\mu) - \exp(-0.050\mu)}{\exp(-0.050\mu)}.$$

Half-value thickness (HVT) = 12 mm, so $\mu = \log_e 2/\text{HVT} = (0.693)/(12 \times 10^3)$ = 57.75 m^{-1},

$$\frac{\Delta N}{N} = \exp(0.003\mu) - 1 = \exp(0.17325) - 1$$

$$\frac{\Delta N}{N} = 0.189 = 18.9\%.$$

6.2

From Table 6.1, the difference in mass absorption coefficients for nickel and titanium is greatest for Cu Kα radiation at a wavelength of $\lambda = 0.154$ nm.

$$\text{Ni: } \mu_m(\text{Ni}) = 4.57 \text{ m}^2 \cdot \text{kg}^{-1}$$

$$\text{Ti: } \mu_m(\text{Ti}) = 20.8 \text{ m}^2 \cdot \text{kg}^{-1}$$

For a completely homogeneous region of the alloy

$$\text{Ni-50\%Ti} \quad \mu_m(\text{Ni-50\%Ti}) = 0.5 \, (\mu_m(\text{Ni}) + \mu_m(\text{Ti})) = 12.7 \text{ m}^2 \cdot \text{kg}^{-1},$$

and the density of this region will be

$$\rho(\text{Ni-50\%Ti}) = 0.5 \, ((\text{Ni}) + (\text{Ti}) \,) = 6700 \text{ kg} \cdot \text{m}^3$$

$$\mu = \rho\mu_m.$$

So for the homogeneous regions of the alloy,

$$\mu(\text{Ni-50\%Ti}) = 8.51 \times 10^4 \text{ m}^{-1},$$

whereas for nickel,

$$\mu(\text{Ni}) = 4.04 \times 10^4 \text{ m}^{-1},$$

and for titanium,

$$\mu(\text{Ti}) = 9.44 \times 10^4 \text{ m}^{-1}.$$

The ratio of transmitted intensities through nickel and titanium is,

$$\frac{I(Ni)}{I(Ti)} = \exp\left((\mu(Ti)-\mu(Ni))x\right) = \exp(5.4) = 221,$$

and the ratio of transmitted intensities through the Ni-50%Ti alloy and titanium is

$$\frac{I(Ni-50\%Ti)}{I(Ti)} = \exp\left((\mu(Ti)-\mu(Ni\text{-}50\%Ti))x\right) = \exp(0.93) = 2.5.$$

6.3

The dose rate at a distance r from a point source with a shielding layer of thickness x is given by

$$D_R(x, r) = D_R(0, 1) \cdot \frac{1}{r^2} \exp(-\mu x).$$

At 1 m, the unshielded dose rate is 200 rem, therefore at 5 m, the unshielded dose rate will be

$$D_R(5) = \frac{D_R(1)}{25} = 8 \text{ rem}.$$

The attenuation coefficient, μ, is related to the half-value thickness, HVT, by

$$\mu = \frac{\log_e 2}{HVT} = 139\ m^{-1}.$$

The required thickness of shield is therefore

$$x = -\frac{1}{\mu} \log_e \left(\frac{D_R(5)}{D_R(1)} r^2 \right) = 0.02\ m.$$

MECHANICAL TESTING METHODS

7.1

In this case, we can use the equation for the nonlinear stress–strain curve

$$\sigma = K \varepsilon^n,$$

and taking logarithms of both sides

$$\log_{10} \sigma = \log_{10} K + n \log_{10} \varepsilon.$$

Therefore, a plot of $\log_{10} \sigma$ against $\log_{10} \varepsilon$ should give a straight line with slope n and intercept $\log_{10}K$. Using true stress and true strain values gives:

$\log_{10}\varepsilon$	$\log_{10}\sigma$
−1.311	8.525
−0.708	8.664
−0.678	8.668
−0.447	8.700
−0.276	8.732
−0.142	8.761
−0.031	8.789
0.038	8.816

A graphical solution yields

$$K = 631 +/- 10\ \text{MPa}$$

$$n = 0.2 +/- 0.02.$$

7.2

The critical crack size, a_c, in this material at a stress of 400 MPa (assuming for simplicity, a semicircular crack) is

$$a_c = \frac{K^2}{\pi \sigma^2}.$$

At $T = -50\,°C$, $K = 50$ MPa $\cdot$ m$^{1/2}$.
At $T = 50\,°C$, $K = 140$ MPa $\cdot$ m$^{1/2}$.
Therefore, the critical crack sizes at these temperatures are:
$T = -50\,°C$, $a_c = 0.005$ m.
$T = 50\,°C$, $a_c = 0.039$ m.
For a margin of safety, it is required to detect a crack that is $\frac{1}{2}$ the critical size
At $T = -50\,°C$, $a_c = 0.0025$ m.
At $T = 50\,°C$, $a_c = 0.020$ m.
The main observation here is that at the lower temperature, where the material is brittle, the crack size that needs to be detected is much smaller than at the higher temperature, where the material is ductile.

7.3

As determined from the table, the midpoint between the upper and lower shelf energies is at an impact energy of 37.5 J. The yield strength is 1500 MPa. The equation that relates Charpy V notch test impact energy, CVN, to fracture toughness, K_{1c}, is

$$K_{1C}^2 = \sigma_y^2 \left(\alpha \frac{CVN}{\sigma_y} - \beta \right) = \sigma_y \alpha CVN - \beta \sigma_y^2.$$

The coefficients in this equation are: $\alpha = 0.625 \times 10^6$ m^{-2} and $\beta = 0.00625$ m.
At 20 °C, the Charpy V notch impact energy is $CVN = 58$ J

$$K_{1C}^2 = (1500 \times 10^6)^2 (0.0242 - 0.00625) = 40.3 \times 10^{15}\ \text{N}^2\text{m}^{-3}$$

$$K_{1c} = 200\ \text{MPa} \cdot \text{m}^{1/2}.$$

At 0 °C the Charpy V notch impact energy is $CVN = 23$ J

$$K_{1C}^2 = (1500 \times 10^6)^2 (0.00958 - 0.00625) = 7.5 \times 10^{15}$$

$$K_{1c} = 87\ \text{MPa} \cdot \text{m}^{1/2}.$$

The critical crack size, a_c, is given by

$$a_c = \frac{K_{1C}^2}{\pi\sigma^2}.$$

At 20°C and 1000-MPa stress

$$a_c = \frac{(200)^2}{\pi(1000\times10^6)^2} = 12.7\times10^{-3} \ m.$$

At 0°C and 1000-MPa stress

$$a_c = \frac{(87)^2}{\pi(1000\times10^6)^2} = 2.3\times10^{-3} \ m.$$

ULTRASONIC TESTING METHODS

8.1

The extent of the near-field region is given by

$$\ell = \frac{D^2}{4\lambda} = \frac{D^2}{4\left(\dfrac{V}{v}\right)},$$

where $d = 25$ mm, $V = 5{,}960$ m.sec^{-1}, $v = 5\times10^6$ Hz,

$$\ell = \frac{(25\times10^{-3})^2}{4\times\left(\dfrac{5{,}960}{5\times10^6}\right)} = 0.131 \ m.$$

The beam width at 1 m from the transducer is

$$w = 2\tan\left(\frac{\alpha}{2}\right),$$

and

$$\frac{\alpha}{2} = Arc\sin\left(\frac{1.22\lambda}{D}\right) = 3.33°,$$

therefore,

$$w = 2 \tan (3.33°) = 0.117 \text{ m}.$$

8.2

The signal level measured on an oscilloscope is the amplitude, not the intensity, and therefore the gain, X, is given by

$$X = 20 \log_{10}\left(\frac{A_f}{A_i}\right) = -10,$$

where A_i is the amplitude of the initial pulse, and A_f is the amplitude of the pulse after reduction by 10 dB. Therefore,

$$A_i = A_f \times 10^{10/20} = 3.16 \times A_f.$$

If the amplitude of the signal is doubled, $A_{new} = 2\,A_i$, then the additional attenuation, X_{inc}, needed to reduce the amplitude of the signal on the oscilloscope is,

$$X_{inc} = 20 \log_{10}\left(\frac{1}{2}\right) = -6.02 \ dB.$$

8.3

The critical angle of incidence, θ_{CL}, whereby the longitudinal wave does not penetrate the metal because the angle of refraction is 90°, is given by Snell's law. Therefore, for all angles of incidence greater than this there will be no longitudinal wave transmitted into the metal.

$$\sin\theta_{CL} = \frac{v_{L1}}{v_{L2}}\sin 90 = \frac{2}{4.5} = 0.444$$

$$\theta_{CL} = 26.4°$$

The critical angle of incidence, θ_{CS}, above which no shear wave is transmitted into the metal is

$$\sin\theta_{CS} = \frac{v_{L1}}{v_{S2}}\sin 90 = \frac{2}{2.5} = 0.8$$

$$\theta_{CS} = 53.1°.$$

Therefore, the angle of incidence of the ultrasound in the wedge "shoe" should lie between these two extreme values. An angle halfway between these may be considered optimum.

ELECTRICAL TESTING METHODS

9.1

The frequency of excitation is 70,000 Hz, so that one cycle takes 14.3×10^{-6} sec. Therefore, 50 cycles takes 714.3×10^{-6} sec. The part must travel 4 mm in this time, so the required velocity is

$$V = \frac{0.004}{714.3 \times 10^{-6}} = 5.6 \text{ m.sec}^{-1}.$$

9.2

The eddy current probe at air point has $R = 20\ \Omega$, and $X = 20\ \Omega$

$$Z = \sqrt{R^2 + X^2} = \sqrt{800} = 28.28\ \Omega.$$

The same probe, when placed on aluminum, has $R = 30\ \Omega$, and $X = 10\ \Omega$

$$Z = \sqrt{R^2 + X^2} = \sqrt{1000} = 31.62\ \Omega.$$

When placed on stainless steel, it has $R = 28\ \Omega$, and $X = 15\ \Omega$

$$Z = \sqrt{R^2 + X^2} = \sqrt{1009} = 31.76\ \Omega.$$

The best way to develop a calibration curve for conductivity is to tabulate and plot a graph of the values of R and X for the different values of conductivity, and use this to interpolate values of conductivity from the graph for test materials of unknown conductivity.

	X	ΔX	R	ΔR
$(10^6\ S \cdot m^{-1})$				
0	20	0	20	0
2	15	−5	28	8
35	10	−10	30	10

The graphs will be useful for conductivities within the range $0 - 35 \times 10^6$ S·m^{-1} (that is, for interpolation). Outside this range the graphs are unlikely to be reliable (that is, for extrapolation).

9.3

The initial resistance without any mutual inductance is $R = 20\ \Omega$, and the initial reactance, $X_L = \omega L$, is 12.56 Ω. The change in resistance, ΔR, and the change in reactance, $\omega \Delta L$, are given by

$$\Delta R = \frac{M_I^2 R_1 \omega^2}{R_1^2 + \omega^2 L_1^2}$$

$$\Delta L = \frac{M_I^2 L_1 \omega^2}{R_1^2 + \omega^2 L_1^2},$$

and at a frequency of 10kHz $R^2 + \omega^2 L^2 = 558$ ohm^2

$$\Delta R = \frac{(100 \times 10^{-6})^2 (62.83 \times 10^3)^2 (20)}{558} = 1.415\ \Omega$$

$$\omega \Delta L = \frac{(100 \times 10^{-6})^2 (62.83 \times 10^3)^2 (12.56)}{558} = 0.884\ \Omega.$$

If the variation of reactance with resistance is plotted in the impedance plane, this gives a circle.

$$(\Delta R)^2 = \left(\frac{M_I^2 \omega^2}{R_1^2 + \omega^2 L_1^2} \right)^2 R_1^2$$

$$(\omega \Delta L)^2 = \left(\frac{M_I^2 \omega^2}{R_1^2 + \omega^2 L_1^2} \right)^2 (\omega L_1)^2$$

So that

$$(\Delta R)^2 + (\omega \Delta L)^2 = \left(M_I^2 \omega^2 \right)^2,$$

where $M_I^2 \omega^2$ is the radius of the circle. Therefore, different values of mutual inductance, M_I, give circles of different radius on the impedance plane.

MAGNETIC TESTING METHODS

10.1

The equation to use for a preformed (rigid) coil is

$$Ni = k\frac{D}{\ell},$$

where $N = 300$ turns, $D = 75$ mm, $\ell = 500$ mm, and $k = 11{,}000$ for half-wave rectified currents.

$$i = \frac{kD}{N} = \frac{(11{,}000)(75)}{(500)(300)} = 5.5 \; amps$$

10.2

Because the current is alternating in a "flexible" coil, the equation to use for this case is

$$i = 7.5\left(10 + \frac{d^2}{40}\right),$$

where $i \npreceq 1000$ A.

$$d = \sqrt{40\left(\frac{i}{7.5} - 10\right)} = \sqrt{40\left(\frac{1000}{7.5} - 10\right)} = \sqrt{4933}$$

The required maximum spacing between the coils is therefore,

$$d = 70.2 \text{ mm}.$$

10.3

The field H in the gap is given by the equation

$$H = \frac{H_a}{1 + N_d\left(\dfrac{\mu_{flaw}}{\mu_{iron}} - 1\right)},$$

so the necessary applied field, H_a, is

$$H_a = \left[1 + N_d\left(\frac{\mu_{flaw}}{\mu_{iron}} - 1\right)\right]\frac{B}{\mu_o},$$

when $B = 1.1$ T, $\mu_{flaw}/\mu_{iron} = 0.01$, and $N_d = 0.5$

$$H_a = 0.505 \frac{B}{\mu_o} = 4.42 \times 10^5 \ A/m.$$

If generated by a long solenoid, the required current is

$$i = \frac{H_a}{n} = \frac{4.42 \times 10^5}{25 \times 10^3} = 17.7 \ amps.$$

RADIOGRAPHIC TESTING METHODS

11.1

Using the equation for geometrical unsharpness, with the original source

$$U_g = F\frac{\ell}{L_o} = 4 \times \frac{25}{675} = 0.148 \ mm.$$

The geometrical unsharpness needs to be the same for the new source; with the new size of source and the same thickness of object, the source-to-object distance must be changed to $L_{o \ after}$,

$$L_{o \ after} = \frac{F\ell}{0.148} = 84.5 \ mm,$$

and, therefore, the source-to-film distance will need to be changed to

$$\ell_{after} = L_{o \ after} + \ell = 84.5 + 25 = 109.5 \ mm.$$

The activity of the new source and the source-to-film distance are different, so the intensity of radiation at the film surface will be different, and to maintain the same exposure, ε, the exposure time will need to be changed. The ratio of intensities will be proportional to the ratio of activities (measured in Curies), and inversely proportional to the square of the distance from source to film,

$$\frac{I_{after}}{I_{before}} = \frac{1}{10}\left(\frac{700}{109.5}\right)^2 = 4.09.$$

So the exposure time for the new source will need to be

$$t_{after} = t_{before} \times \frac{I_{before}}{I_{after}} = 1200 \times \frac{1}{4.09} = 293.6 \ sec.$$

11.2

Radiographic contrast is given by

$$C_s = D_1 - D_2 = G(\log_{10}\varepsilon_1 - \log_{10}\varepsilon_2)$$

$$G = \frac{D_1 - D_2}{(\log_{10}\varepsilon_1 - \log_{10}\varepsilon_2)},$$

and $D_1 = 1.5$, $\log_{10}\varepsilon_1 = 2.45$, $D_2 = 2.5$, $\log_{10}\varepsilon_2 = 3.1$
Therefore,

$$G = \frac{1.0}{0.65} = 1.54.$$

Assuming that the linear approximation can be used over the density range of interest,

$$D_1 - D_2 = G(\log_{10}\varepsilon_1 - \log_{10}\varepsilon_2) = G(\log_{10}I_1 t_1 - \log_{10}I_2 t_2)$$

$$1.3 - 2.5 = 1.54 \times \log_{10}\left(\frac{60}{t} \right).$$

Therefore,

$$\log_{10}\left(\frac{t}{60} \right) = 0.8$$

$$t = 10^{0.8} \times 60 = 379 \text{ sec}.$$

11.3

The exposure, ε, of the film needs to be the same in both cases,

$$\varepsilon = I_1 t_1 = I_2 t_2.$$

And the intensity, I, is in both cases proportional to the activity, A, of the source in Curies, and inversely proportional to the square of the distance, ℓ, from source to film,

$$t_2 = \frac{I_1 t_1}{I_2} = \frac{A_1}{\ell_1^2}\frac{\ell_2^2}{A_2}t_1 = \frac{2.2}{(1.2)^2}\frac{(0.45)^2}{0.9} \times 2.5\,h = 0.86\,h$$

$t_2 = 3094$ sec.

The unsharpness equation is

$$U_g = F\frac{\ell}{L_o},$$

and for the second pipe, with $U_g = 0.2$ mm, $L_0 = 225 - 18.75 = 206.25$ mm and, $\ell = 18.75$ mm, the source size must be

$$F = U_g\frac{L_o}{\ell} = 0.2 \times \frac{206.25}{18.75} = 2.2\ mm.$$

Index